HEARING THE ANCESTORS

BRUCE LOVE

HEARING THE ANCESTORS

Archaeology in the Coachella Valley

La Quinta Historical Society

Paperback: 979-8-9892447-0-6

Library of Congress Control Number: 2024933495

First paperback edition May 2024

Cover design by Leigh McDonald, cover photo by Bruce Love
Designed and typeset by Leigh McDonald in Arno Pro 11/14 and Cheddar Gothic Rough (display)

Printed by Printer Name in the USA.

La Quinta Historical Society
PO Box 1283
La Quinta, CA 92247

www.laquintahistoricalsociety.com

CONTENTS

_____

AUTHOR'S NOTE

On Terminology

R EADERS OF THIS small book may notice that I frequently use the term Indian but rarely use Native American, which might raise a few eyebrows. Personally, I felt quite comfortable using Indian in all the places where it appears in the book, so comfortable that I did not even think about it or second guess it as I was writing.

I want to relay a story from my days at UC Riverside (1990-1993). I was teaching an archaeology field class and we had two members from Pechanga in the class. One day as we were working, Doris Whited (in her 50s) asked me "What's all this about Native American? I've been Indian all my life and now I'm supposed to be Native American?" I answered her, "It's because the word Indian was put on you by the White man." She said "Well, what do you think Native American is!"

After hearing some questions from early readers of this work, I sent out inquiries to some of my Native colleagues, to ask if they wouldn't mind giving their perspectives on the issue. I reached out to five Native friends and got two responses. Myra Masiel (Curator, Pechanga Cultural Resources) reminded me it's always good to reference a specific tribe, such as Luiseño or Cahuilla, when appropriate, "but the general word Indian is fine to use." Then she said something that really made me feel good, "It's important to note that I see you as being comfortable with the word Indian because you have spent many years with Native Communities and have developed those relations with us."

On the Cahuilla side, Torres Martinez Tribal Council member Gary Resvaloso wrote, "Indian would be preferred. I really don't care for the label of Native American. To me it's a generational thing: the old people, our a'avutem like my grandfather Ralph Alamo preferred Indian but they pronounced it Injun."

With these responses, I feel comfortable keeping the term Indian, and hope that this brief foreword helps you, the reader, feel more comfortable with my choice to use the term, too.

HEARING THE ANCESTORS

INTRODUCTION

What Archaeologists Do, Really

ARCHAEOLOGY IS NOT treasure hunting! Archaeology is not treasure hunting! The curse of Indiana Jones has dogged our profession for generations, even before Indiana Jones. We feel maligned that the general public has the wrong idea about archaeology. They think we are treasure hunters, and so we speak from a moral high ground as we try to dispel the notion that we are seekers of gold and glory. But secretly, I'm pretty sure most of us like it when someone refers to us as "a real Indiana Jones."

We're not supposed to like it, but we do, and we smile sheepishly as we halfheartedly explain that our goal is not to find the "golden idol with the ruby eyes," as my archaeology professor at UCLA used to say back in 1976 when I took my first archaeology field class. The purpose of archaeology is to rebuild and reconstruct the past based on the evidence left in the ground by ancient (and not so ancient) peoples. How did the people live? What was it like then? What kind of social organization, trade networks, food gathering, population densities, settlement, and subsistence did they have? Can we see remote past lifeways by carefully excavating and examining the remains the people left behind? That's why archaeology is part of anthropology, the study of humans, and again, *not* treasure hunting.

Thus, for an archaeologist, finding small bits of chipped stone or burned plant remains can have more significance than finding a complete arrowhead or intact mortar and pestle, because the scientific information contained in the tiny bits can be greater than the knowledge gained from a whole pot or unbroken spearhead.

Archaeologists are trained in colleges and universities in archaeological theory and investigative methodology. We learn that one should never touch spade to ground without first developing a research design: we don't go out to dig things up; we go out to answer research questions. Before excavations, we outline the questions we're trying to answer, and we ask what kinds of findings can address those questions. Archaeologists dig for knowledge, not for treasure. At least that's what we learn in the university classrooms.

Then we leave the hallowed halls of academia—except for the 1 or 2 percent who get teaching jobs—and we enter the world of real archaeology, and for most of us that means CRM archaeology, and I'll explain what that is.

CRM stands for cultural resource management, a fancy name for "contract archaeology." This is the archaeology that's done for land developers, to go out ahead of construction projects to make sure they don't bulldoze Indian burial grounds or destroy historic stagecoach stops—environmental impact report–type archaeology. CRM is a profession, an industry really, with hundreds of companies competing for contracts.

From 1990 to 2003, with a PhD in anthropology from UCLA, I did CRM archaeology throughout Riverside and San Bernardino Counties (at one time building my company up to 20 employees), which brought me to the Coachella Valley and introduced me to this unique land, with its desert communities and their planning departments, expanding golf courses, and rapidly spreading housing tracts alongside established Indian reservations with Native American cultures and values, and 100 centuries of archaeological remains in the ground to be protected.

It was a perfect storm of potential conflict ameliorated by occasional collaboration, a prescription for stuttering progress with opposing motives, headbutting intentions, and widely differing objectives. The head of the CRM company finds himself at the center, in some cases a target and in other cases the pivot point where opposing views can mesh and blend and cohabit.

Doing real research, remembering those lofty goals from our university training, trying to answer outstanding questions that someday might make college textbooks, keeping an eye peeled for important data, unearthing finds with consequence, discovering new aspects of the historical past long covered by sands and gravels of shifting landscapes—all this was done under the pressure of encroaching bulldozers, earthmovers, and scrapers; land developers with budgets and timelines; city planners with their own careers and agendas at stake; keenly critical Native American observers; and competitor CRM companies ready to accuse you of unethical or unprofessional practice. These are some facets of practicing archaeology in the Coachella Valley, the subject of this book.

In the following pages, I narrate some of the highlights of my 13 short years of CRM in the Coachella Valley. In 2003, I sold my company and moved to the high desert in northern Los Angeles County (the Palmdale-Lancaster area) and retired from CRM to return to my Mesoamerican studies, the archaeology and culture of the ancient, historical, and present-day Maya. I return now, occasionally, to the Coachella Valley to renew my old friendships with tribal members and to enjoy seeing youngsters who once worked with me as trainee archaeologists now holding responsible and respected positions in tribal government.

Some local people asked me if I might write about my experiences here in the Coachella Valley, and I thought, yes, it's important for the lay public to know more about the archaeology here and some of its more fascinating features. I have agreed to put together a few chapters, for the record, of what we did and what we found during that decade plus of hard work, anxiety, fret, push-pull, self-doubt, self-propulsion, and finally an overall sense of satisfaction and self-approval for efforts well made, certainly well intentioned, and sometimes well completed. If you, the reader, find these tales half as interesting as I do, you've got a good read ahead of you.

FIGURE 1. Rancho La Quinta entrance on Washington Street.

CHAPTER 1

Does Finding Something Mean You Can't Build There?

————————

T HAT'S THE BIG question, isn't it? Developers are understandably afraid of archaeology. They're afraid it will kill their project if you find something, but in fact it almost never does. In 13 years of archaeology in the Coachella Valley, I never killed a project, partly because I didn't have that kind of power and partly because I always tried to find a solution, a common ground, a compromise. One of the best examples of collaboration and cooperation I can think of during that time was the Rancho La Quinta Country Club project at the southeast corner of Washington Street and Avenue 48, some 350 acres of mesquite, creosote bushes, and sand dunes, dry, hot, desolate, and slated to one day be a PGA-class resort, eventually growing to 700 acres.

The City of La Quinta, to its great credit, insisted that the project go through archaeological inspection prior to approval of grading permits and other go-aheads. I say the City of La Quinta, but in fact, it is individuals, people within city government, that make things happen, or members of civic organizations that are watchdogs, that keep a clear eye on government, that do not allow projects like this to slip through the process. The city told Rancho La Quinta at the earliest point, before grading of any kind had happened, that it had to hire a professional archaeological consulting firm to "survey" the property and file its report with the city.

FIGURE 2. Before: typical creosote dune, how the Rancho La Quinta property would have looked in 1997.

FIGURE 3. After: how the Rancho La Quinta sand dunes look in 2023.

PHASE I: SURVEY AND RECORDS SEARCH

"Survey" is a jargon term used by archaeologists. It's not a survey in the sense that land surveyors find corners and flag property lines; this kind of survey is what CRM archaeologists call phase I, or more accurately, one part of phase I. When archaeologists survey, they walk parallel lines back and forth across the property at close intervals to inspect the ground surface and pin-flag artifacts or features if they see any.

NOT DIGGING

The general public thinks archaeology is digging, and it can be digging, but by far the bulk of archaeological fieldwork is survey work, inspection of the surface of the land. Trained and experienced archaeologists will spot things that the untrained person would never see, such as small chips of stone that are the result of Indigenous toolmaking, or a change in soil color that suggests the ground might contain artifacts, or other remnants of past activities.

The survey team, walking parallel transects, pin-flags all findings: you see a pottery sherd (I'm not sure why we say sherd and not shard, but we do), you flag it and move on; you see a flake (chipped stone), you flag it and move on; you see a fire-affected rock, you flag it and move on. After the property is thoroughly surveyed, you look back and see the clusters of pin flags, and those clusters are the sites. A site can be as few as three contiguous artifacts or as many as thousands.

The other part of phase I, which should be done even before the survey work, is the records search. This is the background search for previously recorded sites, previous studies, old maps, and documents of any kind that might throw light on the history of the property. There is a statewide system of information centers that keep these kinds of records, with access limited to certified and approved archaeologists and researchers. For Riverside County, the information center is at the University of California, Riverside (UCR).

I remember going into the construction office at the Rancho La Quinta project, a trailer and little else, where the construction manager interviewed me. "This is a lot of money," he said, looking at my proposal. "If I agree to this, will you personally be here to supervise the archaeology—in other words do I get the A-Team?" "Yes," I assured him. I don't know how many competitors he interviewed for that job or how our proposals compared or how much they were bidding, but then and there he gave me the green light, I got the job and

FIGURE 4. Cluster of flagged artifacts: (*from left*) bleached bone, fire-affected rock, pottery sherd.

did not let on that I secretly had no idea how or where I would find enough archaeologists to survey 350 acres in a reasonable time frame. But I had a window before I got started because I would first do the records search.

At that time, there was a two- or three-week turnaround to get the records search done—at least that is what I let on to the developer so I could have some breathing room as I geared up for the fieldwork. You can go into the information center yourself and do an in-house records search, or you can provide them with a map and have them do it. The maps at that time were hard copies of USGS topographic quadrangles with known sites and previous surveys marked on them. Being the 1990s, digitization was only just beginning, and GIS and GPS were things of the future.

The Rancho La Quinta property showed very little previous archaeological activity; few had gone out before and recorded sites there, and only two formal survey reports on small portions of the property were on file to show what I might expect. We would be covering mostly new ground, and everything we found and recorded was going to expand our current archaeological knowledge of the Coachella Valley. I loved this! Making a real scientific contribution. How little did I know that beyond the scientific value, there was a world of

FIGURE 5. Present-day entrance to the Torres Martinez Indian Reservation.

cultural value waiting to be tapped, the Cahuilla world of the Torres Martinez Indian Reservation.

TRIBAL PARTICIPATION

A few years before this, I had made my first trip down to Torres Martinez. I don't remember the circumstances exactly, but somehow I paired up with a youthful tribal firebrand named George Auclair, who was leading protests at that time against dumping contaminated waste on tribal lands, who was interviewed on the news magazine program *20/20*, who invited United Farm Workers activists to grow the protests, and who was a culture bearer of some renown. To George and many others on the rez, protecting artifacts and sites—especially ancestral burials—was vital and urgent. He took me to sites that he knew, and when we found bones that might possibly be human, he taught me what had to be done. We had to consult the tribal medicine man, the spiritual leader at that time, Mr. Ernest Morreo.

I clearly remember George directing me to Ernie's trailer for the first time, and the Indian etiquette he demonstrated as we patiently waited for Mr. Morreo

FIGURE 6. Historic photograph of Ernest Morreo, tribal spiritual leader.

to come outside, and the quiet narration of why we were there: we had discovered some bones that could be ancestral, and handling them could be very
dangerous for us and for the tribe. Mr. Morreo looked at the bones in George's
hand in his silent and humble, yet forthright, manner. Without hesitation he
motioned for us to follow him as he went around to the back of his parked car
and opened the trunk, exposing his sacred bag with the great eagle feather
emerging from it.

From the bag he took out his special headband, a bandanna imbued with
years of sacred vibration and ritual energy; he tied it around his head. Then
he drew out some native tobacco and some rolling papers, and deftly rolled a
cigarette. I don't remember at that point if George was still holding the bones
or if Mr. Morreo had instructed him to set them down, but what followed was
my first introduction to Cahuilla religion, offering tobacco to the four directions and the sky and the earth.

As George and I watched, Mr. Morreo, facing east, lit a hand-rolled cigarette, drew on it to fill his mouth with tobacco smoke, and then blew out the
smoke in a big puff while at the same time pushing the smoke out and up into
the sky with the eagle feather. He then turned 90 degrees to face the north and
repeated the sacred act, then west, then south, then up to the sky and down
to the earth. At one point, he blew tobacco smoke over the bones themselves,

blessing them, cleansing them. Then he instructed us to take the bones back where we found them, to put them in the ground, to cover them up and leave them be. George and I left Ernie's property that afternoon, knowing we had done the right thing. We felt comfortable and safe. Returning to the site, we put the bones in the ground and walked away.

My involvement with Torres Martinez continued to grow. With new housing tracts and golf courses exploding across the desert, I had lots of CRM jobs and lots of surveys, and thus plenty of opportunities to visit George on the rez, and I began to consider the possibility of using tribal members as survey team members, as archaeologists, to help me in my surveys.

The idea was unconventional because trained archaeologists are supposed to have university degrees, are supposed to go through the academic clearing house of classroom studies, tests and examinations, and approved and certified field schools—but I knew that tribal members like George, and others he introduced me to, could actually spot artifacts much better than the unaccustomed non-Indians, not to mention better take the brutal outdoor exposure to sun and wind and extreme dryness. I knew in my heart that tribal crew members could do a good job, but I was afraid of what other professional archaeologists would think of me, or even what they could do to me, such as file a grievance against me for unprofessional conduct.

Then I struck upon an idea, to offer formal training in archaeology to Torres Martinez tribal members, with a curriculum and a completion certificate. I would do this ad hoc, on my own, without the backing of any academic institution, but this would solve my dilemma: the trainees would get useful and important knowledge, and, if anyone challenged me on that subject, I could defend hiring them because they did have some formal training.

So I taught classes on several days down on the rez and then hired the students as archaeologists to do survey with me. It worked. And boy, did we survey!

Doing the fieldwork. Three hundred fifty acres, 1 mile by 0.5 mile and then some, of bone-dry, steep-sided dunes, mesquite and creosote, no shade. I plotted out some doable pieces of landscape, maybe 40 acres square, and we lined up on one side, pin flags and water bottles in hand, spaced ourselves out about 15 feet apart, set a flag to mark where we would return to, and starting walking in a line. Our eyes swept the ground in front of us, looking for artifacts, which took no time, pottery sherds here, there—"flag it" and call out what you've found so others in the line will know. Keep moving, we'll come back to this spot later for a more thorough inspection.

We crossed our quarter of a quarter section, set flags with long-enough pink ribbons to flutter in the wind, to see from a distance, and paced over to start our return, heading back to the flag we had set at the beginning. Aim for the distant flag to keep in a straight line, but keep looking at the ground to pin artifacts—do both. As we approached the area with the pin flags from our previous transect, we found more sherds, put more flags, expanding the size of the site. In increments, moving across the landscape in parallel lines, back and forth, we covered the first 40 acres.

Besides pottery sherds, we found burned animal bone, fire-affected rock, chipped stones that are the remnants of Indian toolmaking, small pieces of decomposed and fragmentary schist that once were grinding stones, anomalous pieces of burned clay, saltwater shells (also in small fragments), and charcoal bits. Anything that was there because of past human activity constitutes an artifact.

Freshwater shells abound, but they are not considered artifacts because they are a natural element left over from retreating Lake Cahuilla (see chapter 5), but if the shell is seashell, it is an artifact because only humans could have brought it here. Shell was used extensively by local tribes; trade with the coast was constant, with shells showing up in Coachella Valley archaeological sites from the Sea of Cortez as well as the Pacific shoreline.

Small, sun-bleached rodent bones are common, but they are generally considered to be natural, especially small ground-up pieces that come from owl pellets that dry out and scatter in the wind. However, if the bones are burned, darkened by fire, they become artifacts because they are the leftovers from cooking fires or roasting pits, evidence of past human activity to be studied and interpreted by archaeologists.

After finishing our first 40-acre section, we looked back and took in the clusters of pin flags. Those were the sites that we were going to map and record, all part of the phase I survey.

Survey results. The records search told us that some sites had been previously discovered by earlier archaeologists, but the entire 350 acres had not been surveyed before, only a part of it, from which we had two reports and some previous site records to refer to. At the end of three weeks of survey, we had found 39 sites! Does that mean Rancho La Quinta can't build there?

We're jumping ahead in the story. The sites need to be recorded first, all as part of phase I. We can't get into the implications yet, but of course in the minds of the developers, that question is foremost: does finding 39 sites spread across 350 acres mean we can't build there? We'll return to that question shortly, but first we need to record the sites.

Recording sites. Clusters of artifacts don't become sites until they are officially recorded. Tough decisions need to be made regarding site boundaries and lumping vs. splitting. There are some rules of thumb, like the 50-meter rule: if one cluster of artifacts is 50 meters from another cluster, they are two separate sites, but if they are closer than 50 meters, they become two loci within a single site. What if you find two clusters 50 meters apart but then find a single potsherd in between?

It's easy to see how arbitrary this can become, but sites must have boundaries, otherwise how can planners and developers proceed? How can archaeologists proceed? By the time the phase I final report is written and submitted, what looked in the field like random scatters and poorly defined groups end up being discrete entities, quantifiable and manageable, recorded, named, and registered within a statewide inventory of "historical resources."

In our case, on this 350-acre project, we ended up lumping many of the sites together into larger sites, and in the end, we recorded 18 sites instead of our original 39. Many small clusters of artifacts that we originally listed as individual sites became loci or concentrations within the larger boundaries

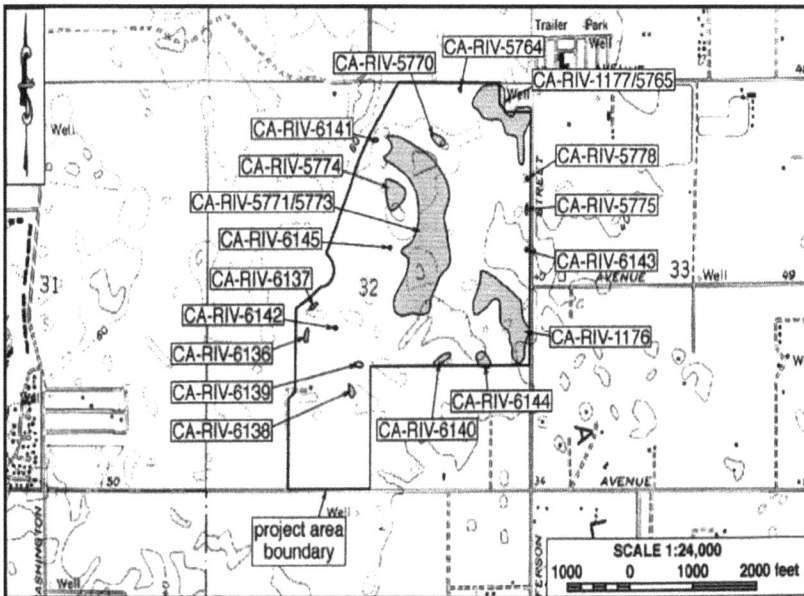

FIGURE 7. Eighteen sites outlined on a map of the Rancho La Quinta project area.

of the extended sites. Like I said, more art than science. As project director, I had to make such decisions based on best judgment and career experience; it is not cut and dry.

As mentioned, when we did the records search, we found that previous archaeologists had surveyed part of the property and recorded sites, and that was because those archaeologists followed professional guidelines and procedures and submitted their findings correctly to create a permanent record for later archaeologists. It was our professional obligation to detail and systematically record all that we found, and that was how we came up with 18 recorded sites in the end.

Returning to the question. Does that mean we can't build there? That's always the hardest question to answer. Personally, it makes me squirm. I tell the developer we need to go to phase II to determine the significance of the sites and that only the "significant" sites will require mitigation. It's going to cost you a boatload more money, and I still can't tell you for certain if you'll be able to build there.

My discomfort comes because what I tell the developer may not be the same thing I would tell a group of professional archaeologists or a group of Native Americans. I feel dishonest because I have a hard time telling the same story to each of the parties, and then of course, there is the Planning Division at the City of La Quinta; they're the ones issuing the permits and approvals. They read my reports and make decisions based on my findings and recommendations (or ignore my recommendations and proceed on their own), and planning staff are subject to the needs and wants of the City Council, who in turn are much more likely to be friendly with the backers and owners of the future golf resort (they don't teach you this in grad school).

I have worked in the nonacademic world, the real world of construction. I even worked in the offshore oil fields in the North Sea and off Long Beach. I've been a roughneck and a rough framer. I sympathize with the commercial world, and I don't share a lot of the more liberal-leaning, environmentally sensitive views of my fellow graduate students and professors. I came to academia late in life and brought my working-class outlook with me, so when a construction superintendent, from his modest desk in a trailer parked out on a future million-dollar golf course, asks me what it means that we found 18 sites on his property, I want to tell him, "Don't worry, we can take care of this for you; this won't kill your project."

But another archaeologist—who might also happen to be a competitor—might say I'm caving to the developer, or worse, I'm in the pocket of the developer, whereas the Indians might say all these finds are significant; they all need to be protected and preserved. How to proceed?

The answer is to go to phase II.

PHASE II: TEST EXCAVATIONS

I mentioned earlier that the whole world thinks of archaeology as digging, but so far in this development project—the Rancho La Quinta golf club and resort—we have completed a thorough surface inspection of the entire property, found 18 sites, and completed phase I of our project, all without touching shovel to sand.

Site significance. Now, to evaluate 18 sites and determine which ones are significant and which ones are not (a hopelessly simplistic dichotomy), we dig, we do controlled excavations. Again, we are not looking for treasure; we're looking for information, for data, for knowledge. Sites that have abundant archaeological resources on the surface and beneath the surface have the potential to fill gaps in knowledge and to build models of how people lived in the ancient past. Sites that contain such information meet the established criteria for significance and therefore require further consideration—perhaps preservation, perhaps not—but certainly some form of mitigation.

Geology and geomorphology. One of the strangest paradoxes in the physical sciences is the disconnect between archaeology and geology. Of course, these two disciplines should be close cousins, with fluid exchange of information, data, and methodologies. Archaeological sites are in the ground, after all, and the ground—its makeup, how it was formed, its layers and levels—is all within the purview of geology. The chipped-stone artifacts are rocks, for goodness' sake, and who knows rocks better than geologists?

Site formation, known as taphonomy, the geologic processes that create the matrix within which the artifacts are found, is so important for archaeologists to understand—was this soil washed in as alluvium, carrying the artifacts with it, or was this ground the actual location where the Indians were doing their activities?

I remember one dig I was on near Vandenberg Air Force Base, in sand dunes just in from the ocean shoreline, monitoring a backhoe operation that came down on a layer of ground-stone artifacts maybe five feet below the surface. Manos and metates and other modified stone artifacts began coming out in the backhoe bucket and sliding down the surface of the backdirt pile as the backhoe operator dumped his buckets of sand.

Why are all these artifacts occurring in a single level, in one stratum, I asked myself, as we collected the artifacts and noted their provenience. It was explained to me later by a geologist who specialized in sand-dune formation. At one time, the artifacts were on the surface, where the Indians were camping. After the site was abandoned and left to Mother Nature, the dunes with the artifacts went through cycles of deflation and buildup. Sometimes, when wind and moisture conditions were just right, the dunes would slowly decrease in height. They would deflate. The sand would move away horizontally and the artifacts, too heavy to be blown sideways, sank along with the surface of the dune itself.

Over untold years, the surface of the dune was lowered several feet (meters) down to a hardpan of soil that underlay the dunes, and all artifacts of course came down with them and lay there in a single layer. Then, over more decades or centuries, new sand came in and covered the layer of artifacts, gradually building up the height of the dune to its present height and shape.

Repeated cycles of dune formation and deflation left what could have been multiple surface occupations all conflated into one layer of heavy stone artifacts along the lower levels of the dune, at its base, set on the hardpan of compact soil. Thus, the artifacts so harshly and unceremoniously grabbed and extracted by the backhoe, could represent multiple surface occupations over centuries, or even millennia, all brought together by geologic processes little understood by most archaeologists.

I personally was enlightened regarding what we saw in the backhoe bucket and backdirt only after the fact, by consulting with a geologist who understood dune morphology, how that site was formed over time by the confluence of human activities and the natural forces that subsequently created the archaeological record: site taphonomy.

An aside. I can't leave my story of the Vandenberg monitoring project without commenting that in those few days, I experienced the coldest July weather I have ever known, bone-chilling winds blowing in off the Pacific in the middle of summer, there around Point Conception.

I also must tell an upsetting story. On that Vandenberg job, I was paired with another archaeologist, and, unfortunately, he showed no respect whatsoever for the artifacts we were finding or the Native Americans who left those artifacts behind, or their living descendants for that matter.

At one point, a soapstone fish effigy appeared in the backdirt, not too finely made, a small oval of soapstone with a tail and a little indent for a mouth. An amateur might have thought it just a rock, but I said to him that it looked like it might be a fish effigy. He kind of grinned and took his pointed trowel and ground its tip into the mouth of the fish, twisting his hand back and forth, drilling the trowel into the artifact, desecrating and destroying it before tossing it aside with a chuckle. I was shocked and emotionally moved, but too cowardly to say anything. It has been a more than unpleasant memory ever since. Perhaps my current work with Native Americans is an effort to make amends for such behavior in the past, when I was a beginner.

Surface scrapes. Different archaeologists have different methods based on their personal training and experience in the field, and also depending on the terrain, the type of ground, the efficacy of certain methodologies over others, and the old question of budgets and personnel. Our standard methodology was to do surface scrapes in areas of high artifact concentrations, which entails scraping across the sand with a square-nosed shovel within an outlined area, two-by-two-meter square, one-by-two-meter rectangle, etc., and screening the sand through ⅛-inch screen, scraping down some 20 or 30 centimeters (roughly 8 to 12 inches) in order to get a good sample of artifacts. Then the vertical excavation unit is often placed within the surface scrape.

Placing the units. The old saw that it's more art than science certainly applies here, but the methods are rather mechanical and straightforward, and a typical test excavation can be described simply. Let's say the original site, as recorded during phase I, is an oval shape about 20 meters long by 15 meters wide consisting of some 35 to 40 artifacts—pottery fragments, chipped stone, fire-affected rock, and ground-stone fragments—that are scattered across a low area between surrounding sand dunes.

We might decide that 10 test units would be adequate to determine whether the site has a subsurface component and, if so, how much of one. Using a compass, a string line, and a measuring tape, we lay out 10 one-meter-square units in a north–south pattern across the 15-meter-wide oval, and outline the 10 small squares, the one-meter-square units, with stakes and string line.

(As an aside, you notice everything is done in metric measure. Well, I am old enough to remember when California archaeology was done in feet and inches. A standard test unit was five feet square, and excavations were done in six-inch levels. At my first field class in 1976, my UCLA archaeology professor—who even then was considered a bit of a dinosaur—was defending the English system against the rising tide of metric. During lunch breaks at the dig, he would go on about the revisionists trying to change everything to metric, which of course, in the long run, they did.)

In the academic classrooms, there are whole chapters and lectures and written papers (and exams) on sampling strategy—in other words, if you decide to excavate 10 units in a site that is 20 by 15 meters, how do you lay out the units to get the most informative results? This is called sampling a site. To dig one-meter-square controlled excavation units is in fact to take a sample of what lies below the surface, and whole dissertations have been written about the pros and cons of different kinds of sampling, all of which have ingrained biases.

Do you put the units only where the greatest concentrations of material are visible on the surface? That would introduce a very strong bias. Do you space out the units evenly across the surface regardless of where the concentrations are? You would then be reducing the bias, but how you space the units introduces other kinds of bias. If you space out the units evenly, that's a bias. If you space them randomly, there's another bias.

There are advocates who propose that unit location should be decided with a random number generator to reduce bias, but even then there is the question, do you place all units within the site boundaries, or do you place some units outside the site boundaries to get comparative sets of data and to see if there are subsurface remains where there are no visible artifacts on the surface?

It's all very academic and actually has little practical value in cultural resource management, where the biggest consideration—which is not taught in school—is budget and manpower, personnel availability, and caps on time and money. But it is good to keep sampling issues in mind, or at least to be aware there are such issues, when placing your 10 units across the site.

I end up putting in 10 two-meter-square surface scrapes with a one-meter-square test unit in the center of each one, laid out in two rows of five test units each, evenly spaced and spanning the oval-shaped site regardless of visible artifact concentrations on the surface.

Excavation. We dig the test units in 10-centimeter levels screened through ⅛-inch screen. Using a trowel and bucket or a square-nosed shovel, the excavators scrape down through the sand inside the string line, marking the one-meter square. As they proceed, they scrape carefully across the surface,

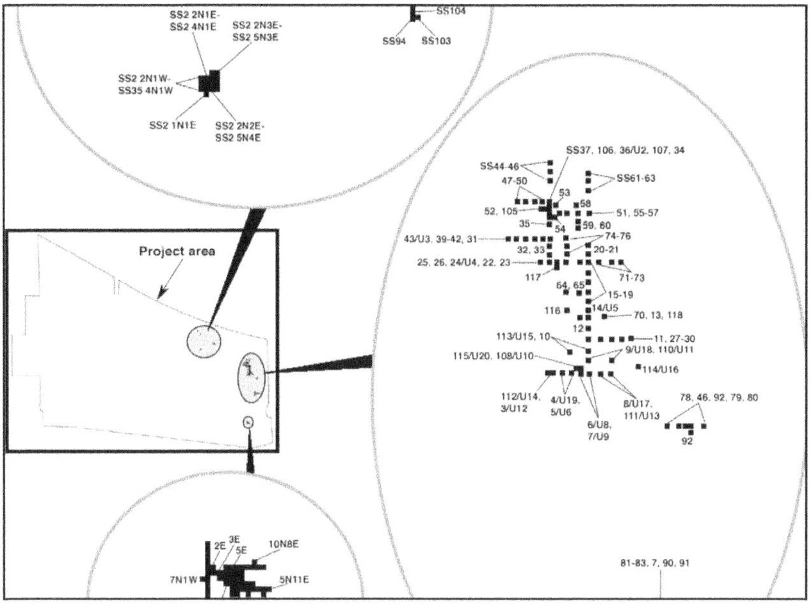

FIGURE 8. Indian Wells Country Club, a very complex site with multiple concentrations, was tested with surface scrapes and test units laid out in various patterns.

not gouging or digging but drawing the trowel or shovel across the sand so that if they hit something, like fire-affected rock or a grinding stone, they hear the scraping sound and feel the bump and can leave it in place.

Measuring down from the string line along all four sides, the excavators make sure the hole is level, 10 centimeters deep all the way across from side to side and in the center. This is the first level. All the loose sand goes into a nearby screen as the diggers clear the hole down to the bottom of the first level, 10 centimeters (about four inches).

The screen is a shallow box with wooden sides and ⅛-inch wire mesh (hardware cloth) stretched across the bottom. The box is attached to legs so the screeners can rock the box back and forth as the sand falls through the mesh. Artifacts like pottery sherds and chipped stone remain in the screen as the screeners screen and the diggers dig, and they are then bagged in labeled level bags, one bag for 0–10 centimeters, one bag for 10–20 centimeters, etc., as work proceeds and the archaeologists take the unit down.

FIGURE 9. Archaeologists screen sand through ⅛-inch hardware mesh, digging in 10-centimeter levels, first 0–10 centimeters, then 10–20, then 20–30, and so on. All artifacts found in the screen are bagged in level bags for later analysis.

Large pieces, as mentioned earlier, like fire-affected rock or ground-stone fragments, are left in place, photographed, and mapped before being taken out. Every level gets a record sheet, a paper form for notes and details about that particular level. If larger objects like whole grinding stones and perhaps clusters of rock are found, they need to be measured and drawn on the level record.

After a dig is completed, it is the level bags and the paper level records, along with photographs, that make up the results of the dig. It is from these that the post-fieldwork laboratory analysis is able to tease out the valuable information, the patterns of data, that are the building blocks of interpretation, analysis, and conclusions about when and how Native Americans were living there, all based on the archaeological record.

Determine significance. If a site turns out to be a small scatter of flakes or sherds on the surface with nothing below the surface, it is usually considered "nonsignificant." If there is substantial surface and subsurface remains, the site is "significant" due to its information value—that is, the site's ability to answer research questions. Sites with human remains immediately rise to "significant" due to the intense interest and concern from the Native American descendants of the very people we are excavating. Having tribal members on the work crew was hugely important in maintaining relations with the neighboring tribes.

Sites that are deep and show extensive occupation over extended periods of time and sites with human remains are significant and require mitigation. Most of the sites turned out to be surface scatters only, and thus it was determined that the work already done was sufficient data recovery and the sites required no further consideration in the planning process, but there were enough "significant" sites at Rancho La Quinta that we needed to recommend phase III: mitigation.

PHASE III: MITIGATION

Phase III means further and more extensive excavations of the deep sites and, regarding the human remains, either removal or preservation in place depending on the recommendation of the Native American consultants. Having tribal members on the crew, as already mentioned, facilitated that consultation.

We were now pushing up against the country club's timeline and their patience. When we started the job, I submitted a proposal for a phase I records search and survey, then I had to break the news to the owners, who were writing my checks after all, that we needed phase II testing to determine site significance, and now, as I stand in the construction trailer, hat in hand, I announce we're not done. We need to go further. The sites that were determined significant required mitigation. Building a country club on a property full of archaeological sites is destructive to those sites and that damage needs to be mitigated, in other words, lessened to a reasonable degree.

One form of mitigation is "data recovery"; another is preservation in place, avoidance. Data recovery means to go in with more excavation units, more extensive surface scrapes, wider and deeper probing and screening to get the maximum amount of data from the site before it's destroyed by the scrapers and dozers. Another proposal, another budget and schedule, more delays before Rancho La Quinta gets its approval from the city to proceed.

Luckily the city was on our side. There were enough activists and citizens who cared deeply about protection and preservation of cultural resources who reviewed our reports and let their views be known to the council and planning department that the developers had to follow along. They would not get their grading permits until the city determined they had done everything reasonable to protect the archaeological resources; so we got our contract for phase III.

We did indeed collect massive amounts of data. I'm prompted by the use of this term, "data," to say something here about the difference between academic archaeology and Native American archaeology. I've always thought this distinction I am about to relate to you summarizes quite succinctly the difference between Indians and non-Native academic archaeologists regarding artifacts. Artifacts, to non-Indians, are data, pieces of scientific information that can be analyzed to tell a story. To Native Americans, they are sacred relics, ancestral remains with medicine—that is, animate power.

For non-Indian academics, to take collections out of laboratories or museums and to put them back in the ground, to rebury them, is to destroy scientific data, akin to book burning. For the Native American, to rebury these artifacts is to preserve them, to return them to where they belong, and to help bring peace to themselves and to their ancestors. What a collision of world views! Destruction of data vs. preservation of sacred objects, a difference that is insurmountable in many cases.

In the case of the human remains we had found, which was a cremation site, Ernie Morreo from Torres Martinez, the same medicine man I had gone to with tribal member George Auclair when George had found remains on the rez some two years earlier, was called out to the site to see the remains.

He said we could take them out of the ground after he blessed them, and we had permission to photograph and measure and describe them in a report, if they would be returned to the tribe after we were through. Of course we agreed, and the country club was relieved they didn't have to redesign their golf course to preserve the remains in place. Mr. Morreo blew tobacco smoke to the four directions and the sky and earth with a hand-rolled native tobacco cigarette and eagle feather as we respectfully took off our hats and turned to face the directions in tandem with him.

And so the three phases of archaeology were completed for Rancho La Quinta Country Club, but we didn't let it go there. There was more. We insisted on monitoring during grading due to the very strong likelihood that archaeological remains would be uncovered and exposed during construction. We recommended it, the city required it, and Rancho La Quinta got their grading permits. Finally, the country club and their financial backers could see a brighter future, and grading could begin.

MONITORING

Monitoring during grading requires a set of skills that is utterly underappreciated and not taught in any school or workshop, yet can result in the most consequential findings of all. The reason? Depth. Scrapers and dozers go down to levels never reached in the first three phases of CRM archaeology, and there may lie the oldest remains, sometimes thousands of years older than the surface remains. But to see the artifacts, to recognize change in soil color or other indicators of past human activities, takes focused attention while monster machines in clouds of dust are churning the surface at high speeds—machines with rubber wheels taller than a human that can run over and crush an archaeologist without feeling a bump.

Operators don't appreciate having to watch out for extra people on the ground, so it is imperative that monitors try to understand what the machine operators are trying to do: they are watching their grade stakes that tell them where and how deep to cut, and they work in tandem with other operators as they push and pull thought the dunes. They really don't like slowing down.

FIGURE 10. Scraper wheels compared to an "ordinary" dump truck wheel.

Safety. There's usually a safety meeting in the field 10 minutes before the start of the day, probably at 6:50. On these construction jobs, all personnel show up early—it's part of the culture; nobody is ever late. If the archaeologist wanders in right at 7:00 or a little after, expressing a lackadaisical attitude that might be entirely acceptable on the university campus, it is anathema to construction culture. That archaeologist has lost respect from the operators.

Our lead monitor on the Rancho La Quinta project was Harry Quinn, geologist, paleontologist, and archaeologist. I mention him here because not only did he arrive early and establish rapport with the operators, he brought donuts! Genius!

We had the sense we were all on the same team, the machine operators and the monitors. In fact, in more than one case, an operator would report seeing something in the sand and avoid that spot until we had a chance to inspect it. The general practice for the monitor is to stand to one side as a pair of scrapers pass by and then to fall in behind them to inspect the newly exposed surface. The problem is that the next set of scrapers may be coming up behind you, so you don't have much time, and they don't have horns. You better be watching over your shoulder while eyeballing the ground. When you do see the next scraper approaching, make eye contact with the operator so he knows that you know he is there. That's the most important safety trick: make eye contact with the operator.

What we found. Thanks to Harry Quinn's expertise, derived from a lifetime of geological field experience and dedication to the job, we found buried sites dating to greater than 1,000 years old, and the clue that gave them away was what was not there. Hard to see a negative, but that's what Harry did and what Harry taught us. As the scrapers exposed surfaces dotted with typical artifacts—fire-affected rock, ground-stone fragments, ashy soil, and burned animal bone—Harry noticed what wasn't there, what was missing: pottery.

All the surface sites and the sites we hand-excavated had ceramic sherds, all of them, fractured and fragmented pottery that once were Cahuilla plates, bowls, utensils, and pots. Here, discovered while monitoring during grading, sometimes 10 feet below the surface, were sites with no pottery. Harry Quinn had single-handedly discovered what's known as the Archaic period, also known as the preceramic period, a time before the introduction of pottery to the peoples of the Coachella Valley, a time greater than 1,000 years ago.

At first it was just a suspicion that these sites without pottery represented ancient times when life was markedly different, different at least in the use of

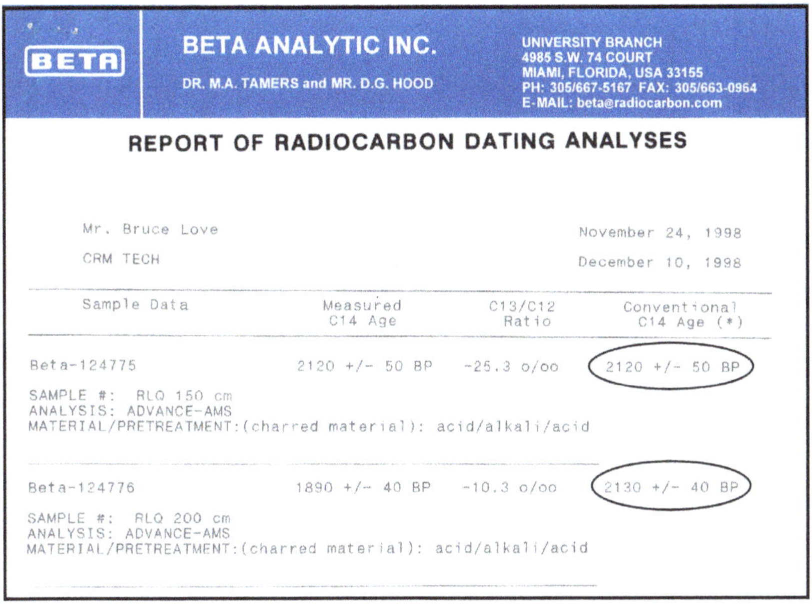

BETA ANALYTIC INC.
DR. M.A. TAMERS and MR. D.G. HOOD
UNIVERSITY BRANCH
4985 S.W. 74 COURT
MIAMI, FLORIDA, USA 33155
PH: 305/667-5167 FAX: 305/663-0964
E-MAIL: beta@radiocarbon.com

REPORT OF RADIOCARBON DATING ANALYSES

Mr. Bruce Love — November 24, 1998
CRM TECH — December 10, 1998

Sample Data	Measured C14 Age	C13/C12 Ratio	Conventional C14 Age (*)
Beta-124775	2120 +/- 50 BP	-25.3 o/oo	2120 +/- 50 BP
Beta-124776	1890 +/- 40 BP	-10.3 o/oo	2130 +/- 40 BP

Beta-124775
SAMPLE #: RLQ 150 cm
ANALYSIS: ADVANCE-AMS
MATERIAL/PRETREATMENT:(charred material): acid/alkali/acid

Beta-124776
SAMPLE #: RLQ 200 cm
ANALYSIS: ADVANCE-AMS
MATERIAL/PRETREATMENT:(charred material): acid/alkali/acid

FIGURE 11. C14 dates from two buried sites on the Rancho La Quinta property, 2,120 +/– and 2,130 +/– 40 years BP (before present).

everyday utensils and vessels, a period before pottery. The deepness of the sites alone suggested great time depth, but to confirm our hunches scientifically, we took charcoal samples and sent them off to a C14 laboratory to be tested, and the radiocarbon dates came back even older than expected: some of them 2,000 years plus!

Here at Rancho La Quinta in 1998 was the first discovery, as far as I know, of the Archaic period (preceramic times) in the Coachella Valley, but once Harry taught us how to recognize them, we at CRM TECH (my company) began to find more of them, especially in La Quinta and nearby Indian Wells. Over the next four years, from 1998 to 2002, we documented no fewer than 12 Archaic-period sites in the Coachella Valley.

I mentioned in the introduction that we university-trained archaeologists try to keep academic research questions in mind while doing CRM. There is such temptation to do the minimum work required, to meet deadlines under budget and move on to the next job, but after finding site after site greater than 1,000 years old—one greater than 3,000!—I felt the need, the weight of obligation, to publish these findings in an academic journal; otherwise this data would lie hidden forever in CRM reports in file cabinets.

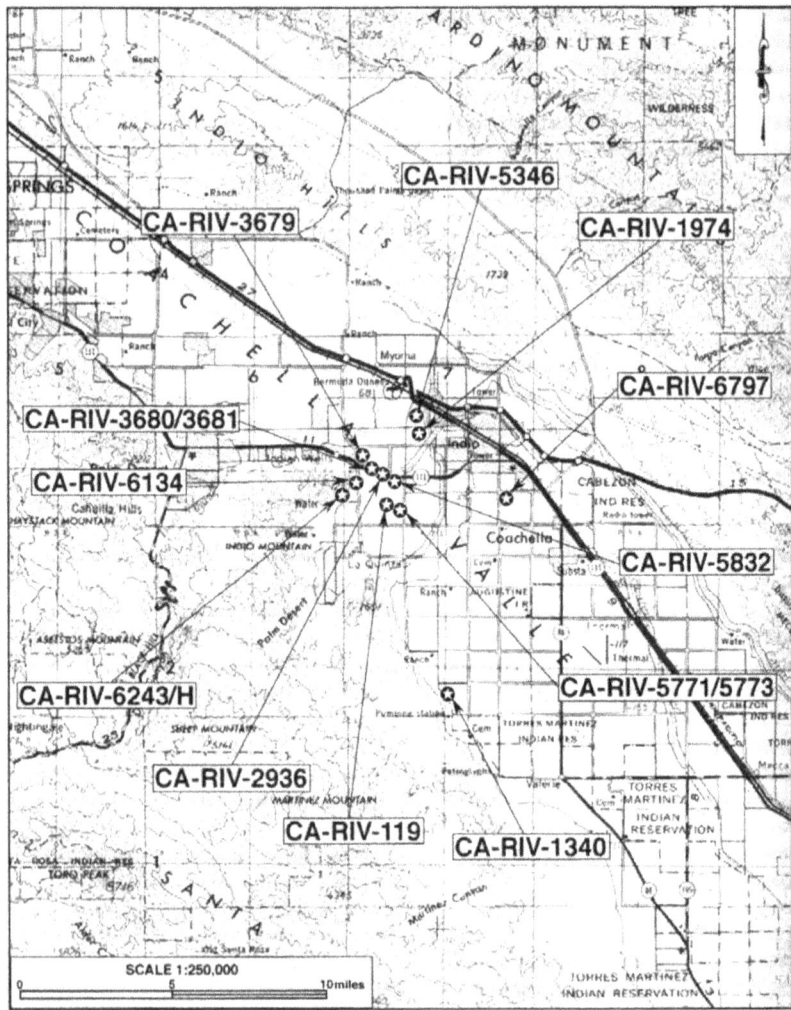

FIGURE 12. Twelve Archaic-period sites discovered in the Coachella Valley, mostly while monitoring during grading. Usually we don't publish exact locations of sites, to prevent pothunters and looters from getting to them, but every one of these sites is now either completely obliterated by grading or lies buried under concrete and asphalt.

Desert Chronologies and the Archaic Period in the Coachella Valley

Bruce Love and Mariam Dahdul

Introduction

Within the last decade, the discovery of more than a dozen sites or features in the Coachella Valley dating to greater than 1000 years allows for a revised overview of Archaic Period life in this portion of the Colorado Desert. A wide range of environmental set-

the eastern Transverse Ranges, occupies the northwestern portion of the Colorado Desert geomorphic province, which, from the Coachella Valley extends southeastward through the Imperial Valley and into Mexico (Jenkins 1980:40-41).

One of the major features to be found within the Colo-

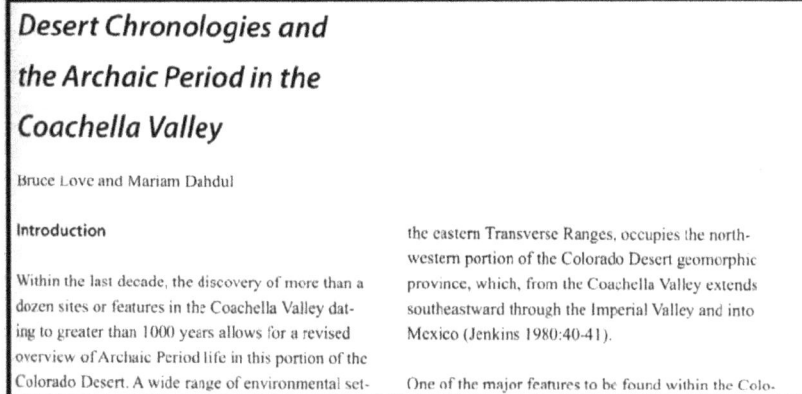

FIGURE 13. An academic article in a professional journal, *Pacific Coast Archaeological Society Quarterly*, now relates for posterity and the archaeological community a summary of our findings: a dozen sites dating from 1,000 to 3,000 years old.

So I enlisted one of my top employees, Mariam Dahdul, to co-write an article with me, "Desert Chronologies and the Archaic Period in the Coachella Valley," which we then published in *Pacific Coast Archaeological Society Quarterly* in 2002. (Mariam has since gone on to get her PhD and is now an important archaeologist for Caltrans, overseeing the protection and preservation of sites threatened by California highway construction.)

The Rancho La Quinta project is an excellent example of CRM archaeology done to good effect: beginning with submitting a proposal to do a records search and survey, getting the job, finding multiple sites that require testing, moving from phase I to phase II, doing phase III large-scale excavations, and finally monitoring during grading.

Typically, this marks the end of a job, the project gets built, CRM reports go into a drawer, and the archaeologists move on, but in this case the findings during monitoring led to a real archaeological breakthrough—the documentation of a previously unknown (to archaeologists) older period of Native occupation. (The Indians will tell you they have always been here, since the beginning of time, but the archaeologists are just beginning to document it scientifically.) And happily, the Archaic-period findings are now in the academic literature, taking their place along with other discoveries, grist for the intellectual mills that grind out models and theories about ancient Native lifeways.

CHAPTER 2

Ancestral Remains Everywhere

M Y FRIEND SEAN MILANOVICH, whose father was chairman of the Agua Caliente Band of Cahuilla Indians for more than 30 years, and who recently (2022) earned his PhD in history from UCR, likes to remind us that "we were everywhere, and our remains are everywhere." It's true.

Everywhere I have worked, from the high desert to low, from the mountains to the coastal plains, if you dig, you find human remains, and why not? Millions of Native Californians lived here for at least 12,000 years; the Indians say much longer. Upon death they were either interred or cremated, depending on tribal traditions, so there is strong likelihood that just about any major ground disturbance, such as grading for construction projects, will disrupt and expose remains.

That's why tribes insist on having monitors present during grading operations. While archaeologists look for artifacts and data, tribal members look for age-old connections, voices from the past, signs of their ancestors' passing. Not just bones but also the items that were buried or burned with them. When Indians say "everything is sacred," that's the last thing archaeologists want to hear—it's a threat to their whole profession; it opens a seemingly irreconcilable gulf between archaeologist and Native American.

My own education along these lines came first from Pechanga, the Luiseño reservation near Temecula. I had arrived at UCR in 1990, fresh from two years running my own small CRM company in the Antelope Valley (Palmdale-Lancaster), where I was not aware of any Native American presence. I was clueless regarding Indian consultation.

UCLA's indecent desecration of a tribal cemetery. Let me back up here to talk about my very first archaeology class at UCLA in 1975. The professor there was a renowned California archaeologist, a senior member of the Anthropology Department faculty, published and respected. I joined his field class in the spring of that year, meeting every Saturday morning for 15 weeks with perhaps 15 or 20 graduates and undergraduates seeking training in how to do archaeology, how to dig carefully and scientifically.

It is almost inconceivable, unthinkable, where we dug. Under this esteemed professor's expert tutelage and guidance, we excavated a known Chumash cemetery in Malibu, California. Prior to our start date, he had a backhoe remove a few feet of recently deposited topsoil in order to arrive at the old ground surface where he knew ancient burials lay. We worked in pairs with trowels and buckets, carefully scraping and screening our excavation units, bagging the artifacts as they came out in the screens, until we reached the depth of the skeletons themselves, mostly decomposed but with many pieces identifiably intact: long bones and skulls, teeth and fingers and toes.

I vividly remember one of the students calling out to the professor, standing up and extending his hands holding each end of a long, exquisitely manufactured chipped-stone blade maybe eight inches long, reflecting sunlight in my memory, bringing *oohs* and *aahs* from the students. It was clearly a grave good buried intentionally with the corpse, an item once possessed by the person now interred, perhaps ceremonially gifted to him, a reflection of power

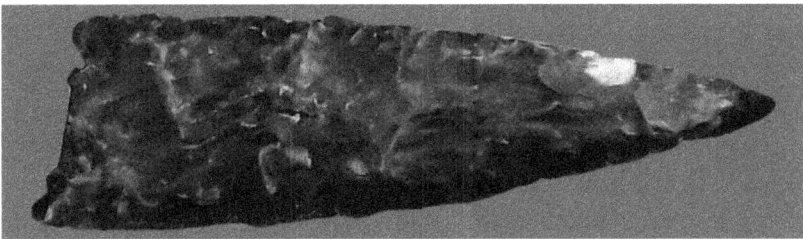

FIGURE 14. Chipped-stone knife blade similar to the one found in the Chumash cemetery in Malibu in 1975. This example was made by a modern flint knapper and is for sale online (https://www.etsy.com/listing/1143202202/genuine-antler-knives-stone-knives).

and prestige and honor. When it was placed in that grave some centuries previous, the mourners expected that sacred stone blade to be buried forever, never again to see daylight, a sacred possession of the deceased, yet its destiny was a drawer in a lab at UCLA.

As troubling as this scenario, and equally disturbing, is that neither I nor any of the students thought twice about the ethics, the immorality, the indecency of our work. I recall during one of our lunch breaks that our professor commented on "fake Indians" protesting the work we were doing, but I never saw or heard of any protestors and did not give it a second thought. This was 1975.

In 1984, as a graduate student in anthropology, still at UCLA, I began working as a crew person on scattered CRM jobs (defined in chapter 1), and I had the pleasure of meeting and working with Kawaiisu culture bearer Harold Williams on a job in Sand Canyon near Tehachapi, but I worked on many other archaeological surveys that had no Native American input whatsoever. In 1987, I started my own CRM company in the Antelope Valley and for the next three years did dozens of surveys and excavations without giving Native American participation a second thought. Like I said, I was clueless. Until I moved to Riverside.

EDUCATED BY THE INDIANS

I was offered the job of director of the Archaeological Research Unit, a part of the Anthropology Department at UCR, essentially to do CRM archaeology for the university, using students as crew people to give them experience and training. I had this innate desire to work with Native Americans and to do the right thing in that regard, but I had no idea how to do it or what I was in for. When I got a survey job near Temecula, I thought it would be the right thing to contact the Pechanga Reservation, so I did, and they told me, "You need to come down here for a meeting."

In 1990, Pechanga had a cultural committee dedicated to preserving ancestral sites, but they had no money and got no respect from the City of Temecula or from contract archaeologists; these were pre-casino days. The only building on the rez was a modest frame community hall and kitchen, an all-purpose place for tribal assemblies, senior luncheons, and cultural committee meetings. I arrived, wide eyed, having never been on a reservation before, stepped into that hall, and got a cultural pounding, a painful earful, thus beginning my education on Indian–archaeologist relations.

FIGURE 15. Raymond Basquez Sr. (Luiseño) in front of the Pechanga Great Oak, singing Naquanish songs with his turtle-shell rattle (https://www.youtube.com/watch?v=7CJA4ypUɪLs).

They could see I was well meaning but naïve. With my PhD and several years of archaeological experience, I couldn't help but believe I knew more than they did. They couldn't even pronounce the words properly, the jargon terminology that graduate students in the university use with aplomb, almost second nature: stratigraphic integrity, lithic debitage, multiple loci, site trinomials, spatial and temporal contexts, mitigation by data recovery, and so on. But I was willing to listen and willing to learn.

My principal teacher from Pechanga was Raymond Basquez Sr., a traditionalist, a culture bearer, a singer of sacred songs, an elder of some distinction. There is not room here to tell of all our meetings, field trips, lunches, and time together, but Raymond was so generous with his knowledge about Indian country and Indian history, the creation of the first peoples, the cremation of their deities, their burial and ceremonial traditions, and what it means to find artifacts in the ground.

ALL-NIGHT WAKES

Around that time, I attended my first Indian all-night wake in the old schoolhouse at Pechanga. Singers with turtle-shell rattles, at that time led by Mark Macarro (later to become tribal chairman), sat in a row in front of the coffin and sang song after song, breaking for a meal at midnight, and then continuing on till dawn, never repeating the same song, telling of the creation of the animals and the people and tribe's migrations in ancient times. There was one other White person there and I remember him confiding to me during the all-night wake, "I feel like I'm in *National Geographic* or something,

FIGURE 16. Anthony "Biff" Andreas (Desert Cahuilla), photo (1996) from the cover of "California Indians & Archaeology," a special report prepared by the editors and staff of *News from Native California*, a quarterly magazine devoted to the history and ongoing culture of California Indians.

watching some tribal ritual." It was indeed a separate reality that I was not aware existed.

On the desert side of the mountains, in Cahuilla territory, Anthony "Biff" Andreas was my teacher. When I got a CRM job in the Coachella Valley and I wanted Native American consultation, I would pick up Biff at his home in Banning and drive down to the desert, show him the property, and ask him his opinion: "Is this a place likely to have been used by the Indians; do you think there are sites here?" I would jot down his comments and include them in my reports. But even more valuable and eye-opening were our off-the-record talks going to and from the sites. Like Raymond from Pechanga, Biff from Agua Caliente was generous to the extreme, telling me stories and facts and history that most cultural anthropologists would die for, but I wasn't doing anthropology, I wasn't extracting data; I was listening and becoming educated in Native ways.

When Biff died in 2009, one of the greatest Native American cultural gatherings in recent Southern California history occurred, the all-night wake and morning cremation of one of the most respected culture heroes in Native Southern California. On the Andreas Ranch, just two miles due south of downtown Palm Springs, singers and mourners came from all over Southern California, from San Diego County and as far east as the Colorado River, singing all night long in a specially built ramada while the wake singers, led by Raymond Basquez Sr. from Pechanga, sang the wake songs in front of the casket.

At daylight, a grand procession with hundreds of singers brought the body out and placed it on a huge funeral pyre of desert branches and logs, a ceremonial leader lit the fire, and as the flames grew higher and reached a crescendo of crackling and sparking, individuals approached the fire and tossed in their precious rattles to be consumed with the body, the greatest honor possible.

Biff's middle name was Lucero, which means "morning star" in Spanish. As the flames reached their peak and began to die down, I looked up in the eastern sky and was pierced by the rays of Venus, the morning star in brilliant display, as if Biff indeed had just risen to the sky and taken his place there.

ALL-NIGHT WAKES AND ARCHAEOLOGY

As I began to attend these all-night wakes, my archaeological sensitivities began to shape up, mainly because these all-night ceremonies were the clearest indication to me that Indian culture was real and strong and ongoing. I remember my UCLA professor talking about fake Indians, and I thought if only people knew about these all-night wakes, they would think differently. But what really caught my attention regarding archaeology and Native culture was the burning of personal items.

In addition to the wakes that were done at the tribal halls, the deceased's personal items were burned at the person's home, in a pit, and then a year later on the anniversary of the person's death, more items were burned, items that had been collected and saved for that purpose since the original burning.

I asked Raymond about this: "Is this is still going on?" "Yes sir," he would reply. Biff told me about one Cahuilla case where the deceased's entire trailer was buried in a huge pit and burned. So I had to think, if archaeologists find burned items in the ground, how do they know they weren't placed there as part of a burning ceremony?

UTILITARIAN VS. SACRED

The perfect example: bone awls. Archaeologists sometimes find these while excavating units at sites, and they are always considered to be ordinary utilitarian tools, usually described as being for basket weaving or leather working. Thin ones are sometimes thought to be hairpins. But when I was on a NAGPRA visit to New York with members of the Pechanga Cultural Committee, I learned different.

NAGPRA (the Native American Graves Protection and Repatriation Act), passed in 1990, is the federal law that requires museums and institutions to inventory their collections and notify relevant tribes of any human remains or associated grave goods. Pechanga was notified by the Heye Foundation Museum of the American Indian in New York that it had collections directly associated with Luiseño culture that had been collected in the early 20th century. Raymond and I and three other members of the Pechanga Cultural Committee flew to New York to inspect the holdings.

They showed us one drawer that had a leather medicine bundle with beadwork opened to expose its contents: sacred stones, shell, stone points, feathers, and a bone awl! An instrument for healing imbued with sacred power, medicine. I motioned Raymond over to look at it, and as soon as he looked in the open drawer, he jerked back and turned his head—too much! too powerful!

FIGURE 17. Bone awl similar to ones found in archaeological contexts and the one found in the medicine bundle at the Heye Foundation's Museum of the American Indian in New York, today part of the Smithsonian Institution (https://www.etsy.com/listing/920064332/prehistoric-indian-bone-awl-artifacts).

dangerous to even look at, much less to handle! Close the drawer quickly and respectfully. Breathe again.

It seems everything archaeologists find is utilitarian, simple household tools or hunting and gathering implements, an interpretation that allows archaeologists to think of artifacts as data, secular and mundane, with no particular spiritual or cultural meaning, just tools and objects. But if you think of the burn pits where personal items are burned during funeral ceremonies and on anniversaries, and then you think of the dark, ashy black middens that archaeologists find that are full of artifacts, including bone awls, you have to reconsider your entire viewpoint on archaeological findings. Reconsidering one's entire viewpoint on anything is almost impossible, though. It is painful and humbling, a process none of us wants to go through.

SACRED OBJECTS FROM THE COACHELLA VALLEY

Viewed with this new perspective on what is sacred and what is mundane, we can look at some of the objects found during my 13 years of archaeology in the Coachella Valley: clay pipes used by medicine men and spiritual guides for ceremonial use of tobacco; stone points not necessarily used for hunting, fastened to the end of ceremonial wands or standalone power objects used in healing and kept in a ceremonial bundle; stone pestles for pounding seeds and nuts in deep wooden mortars, also used to prepare medicine, the private possession of important women such as curers and midwives; cordage and netting used in food collecting and carrying personal possessions, for hanging items inside the house and for wrapping bodies for cremation; basketry, the product of painstaking collecting and weaving based on generations of sacred knowledge, items treasured by their owners; bone awls as sacred implements for healing and curing used by medicine men and women, kept in sacred bundles.

Notice that all the items in this inventory are burned, not a condition you would expect if these were objects simply used and discarded or left behind, facile explanations that make them ordinary instead of sacred. The fact alone that they are burned should be a red flag to the archaeologist that they may very well have been intentionally burned—that is, placed with their owners in a cremation fire (remember the singers throwing their rattles into Biff Andreas's cremation fire) or added to a burn pit of the deceased's personal possessions, a practice still done today at Indian funerals, unbeknownst to the general public and unfortunately unbeknownst to most archaeologists.

FIGURE 18. A: Clay tobacco pipe with handle for holding while smoking and hole in handle for cord or leather strip for hanging, perhaps worn around the neck during important ceremonies. B: Stone smoking tube for tobacco, probably decorated with something tied around the groove near the front. C: Clay pipe with hole in handle, similar to A but showing a different configuration.

FIGURE 19. A ceremonial stone knife, not a utilitarian hunting instrument, reminiscent of the eight-inch blade found in the Chumash cemetery in Malibu in 1975. This one is burned, which indicates it was probably part of a funeral rite.

FIGURE 20. Stone pestle shaped from imported schist, a soft stone probably from the not-too-distant source in the mountains near Chiriaco Summit. This probably was the personal possession of an important woman. The fact that it is burned makes it a sacred object once associated with the deceased's grave site or burn pit for her personal possessions.

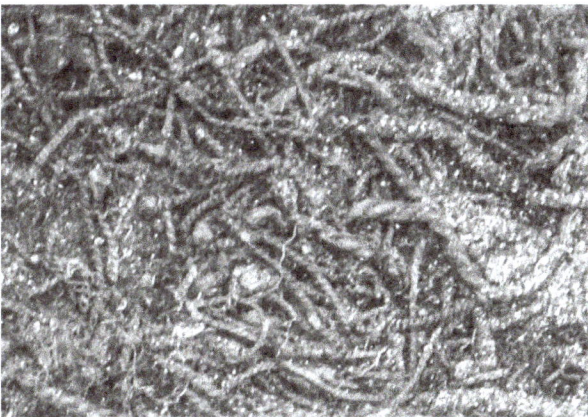

FIGURE 21. Cordage or netting, wrapping for special items, personal property of its Desert Cahuilla owner, burned.

FIGURE 22. Burned basketry. Human cremations often were accompanied by baskets once full of special items burned with the deceased. Infants who died were sometimes wrapped in netting and place in baskets before burning. Baskets were the lifelong possessions of persons, passed down to future generations, or were gifted to others on special occasions.

FIGURE 23. Bone awl. A "tool" thought by archaeologists to be utilitarian in nature, used in basketmaking, but basketmaking itself was a sacred occupation from collecting to weaving, imbued with song and prayer.

FIGURE 24. Bone awl. A particularly thin example, leading some archaeologists to suggest they were hairpins, again utilitarian in nature, not sacred. This one, being very thin, is likely an implement used by a medicine man or woman for curing and may have been part of a medicine bundle.

CHAPTER 3

Rarest of Finds:
A Complete House and Its Occupants, Burned

W HILE MONITORING GRADING at Indian Wells Country Club East Course in 2000–2001, some 4 feet below the creosote bush–topped sand dunes, scrapers exposed potentially the rarest and most consequential archaeological discovery in the history of Coachella Valley excavations: a complete household with remains of the inhabitants and their possessions arranged in a death tableau. These included exquisite tools, the finest products of master craftspeople, ceremonial instruments, mystery objects of marble and mineral, and multiple plant remains, all from sometime in the 18th century, earlier than the first arrival of non-Natives, but still within historic times and well within oral historical times.

It started out as a black sandy smudge. Was it luck? When these massive scrapers come through, they often gouge out swaths 2 to 3 feet deep, 8 feet wide, 30 to 40 feet long, removing massive numbers of cubic yards of sand with each pass, pushing and pulling with the enormous blades on their mid-positioned buckets lowered to cut into and swallow whatever is in its path.

Since the house remains in question are only about 6 inches thick, what are the chances that the scrapers would pass just barely over the top, exposing the ashy change in soil color but not disturbing the feature itself. If the bucket had been lower by 12 inches, the entire feature would have disappeared into the gullet of this mechanical monster, later to be scattered and strewn at some distant part of the property, probably on an ever-expanding hill known as a stockpile, where the scrapers crawl up on one side loaded to overflowing, unload their bellies, and roar down the other side empty, the operators bouncing wildly, turning to come back to make another pass, to take out more sand.

With this scenario playing out pass after pass, scraper after scraper, it seems almost miraculous that our monitor at the time detected the telltale darkened sand of a burned feature before it was destroyed (one cannot help wondering how many features we have lost due to bad luck or less-than-observant monitoring).

I don't recall who made the find that day in the year 2000, but good on him or her. A gray ashy discoloration in the sand alerted the monitor to flag off the area and tell the scrapers to go around while we, the archaeologists, examined the finds.

Careful exposure. When most people think of archaeology, they think of painful patience and delicate care, excavation by dental pick and small brush, miniscule bits and pieces carefully analyzed through magnifying glasses, and so on. CRM archaeology generally means shovels and screens, not bent-over scientists peering into findings inches from the ground. But in this case, as soon as we began to reveal what was at hand, the *National Geographic* stereotype did come into play, down on our hands and knees, gently brushing sand into a dustpan, little by little exposing what had been buried centuries earlier.

The outside edge of the burn area became clear, the dark gray-brown sand, discolored by intense heat, turned suddenly light gray and whitish, the color of natural unaffected sand, the surrounding matrix within which was located this extraordinary, compact collection of burned remains that began to come into focus inside its own clear-cut boundaries as we brushed sand into dustpans and dumped dustpans into buckets to be carried off.

The final shape, as it revealed itself under brushing, was confounding, unexpected, a kind of paisley shape or kidney-bean shape, not round or oval as one would expect. I still don't understand it, but that's what it is, roughly 2.7 meters long (almost 9 feet) by 1.4 meters (4.5 feet) at its widest.

A white mass. Centered in the northwestern end, rising slightly above its surroundings, some 12 inches across, was a mass of something white: rounded, uneven, decomposing around the edges turning the nearby sand whitish, a block of some kind of mineral not easy to identify at first, but which later, upon geological analysis, turned out to be a fine- to medium-grained white marble.

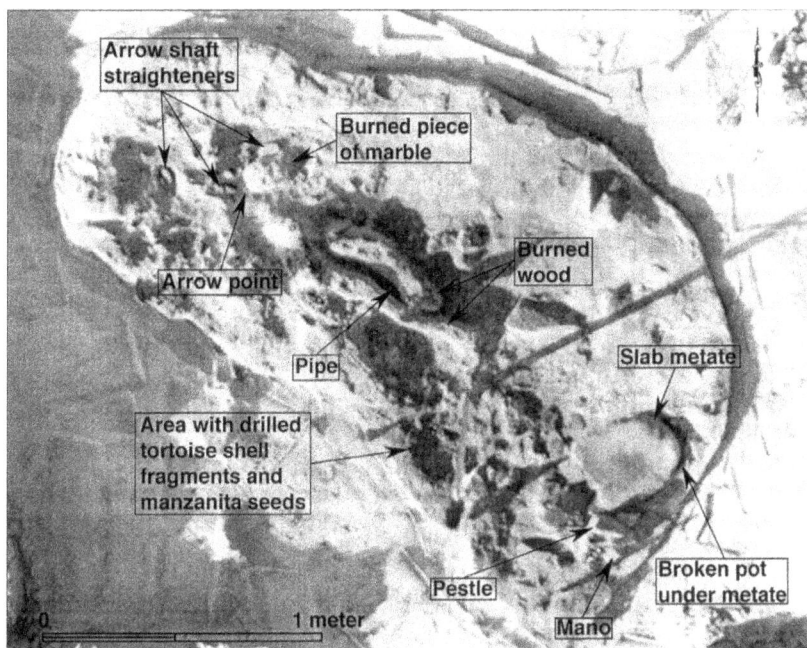

FIGURE 25. Aerial view of burned house, with major features labeled.

If its original shape and surface were known, perhaps we would understand its function, but the high heat from the blaze that once consumed this feature had broken down the marble cobble, obliterating clues that its shaped and worked surface presumably would have provided. Such an item was, and still is, unknown to us, having no counterpart in any excavation or monitoring that any of our team had seen before.

Arrow shaft straighteners. Next to the marble block lay three exquisite arrow shaft straighteners, stones used by arrow makers to work out the bends in the long reeds or hardwood sticks that would eventually be arrow shafts. Nature does not make them naturally straight, though some are very close. Only the experienced arrow maker, having learned from his elders, can select the perfect stone, shape it and groove it, and create an instrument that when heated to just the right temperature and then passed back and forth over the shaft will allow the arrow maker, patiently spinning the shaft, eyeing it, bending it, again and again, reheating the stone, just as his father and uncles taught him, to finally create a perfectly straight shaft that conforms to predetermined standards of excellence, ready to attach points and fletching.

FIGURE 26. White fine-grained marble mass in northwestern end of house.

FIGURE 27. Three arrow shaft straighteners were found next to the white marble.

FIGURE 28. Arrow shaft straightener with incised cross-hatching.

Three such straighteners were found here next to the marble block, one with incised cross-hatching. The decorative cross-hatching tells us this stone was more than simply utilitarian: it went beyond the basic need, it was elevated to a higher plane. Each carefully incised cut on the stone's surface imbued it with its maker's energy and spirit, bringing it alive.

Arrowpoints. Collectors and amateurs say arrowheads; archaeologists say "projectile points," perhaps to set themselves apart. Whatever you call them, one of the most remarkable assemblage of points ever discovered in a single archaeological feature was found here below what is now the Indian Wells Country Club East Golf Course. In this burned house, in its northwestern end, where the white marble slab and the arrow shaft straighteners were found, no fewer than 30 chipped-stone arrowpoints were uncovered within a mere square meter of space, an unheard-of find.

They weren't visible at first, being small and buried, but one or two were seen after the initial clearing exposed the top of the burn. Later as the house remains were swept and sifted through ⅛-inch screen, point after point came out in the brushing and screening process.

Points are generally what archaeologists call "diagnostic," meaning that they signify a certain style and time period. A generic everyday bifacially worked point with no particular defining features might be called nondiagnostic by archaeologists, but the points found here in this burned house were definitely diagnostic; they fit into two categories, both from the late (most recent) preinvasion time period. The two types are Desert Side-Notched and Cottonwood Triangular, as seen in figures 29 and 30.

The Desert Side-Notched points are named, obviously, for the notches in their sides (they also have notches in their bases), and are found throughout the western Unitied States, mainly in the Great Basin (that massive landform between the Sierras of California and the Rockies of Colorado), much of California and most of the Southwest, and are so standard in their morphology (shape), that they have acquired the acronym DSNs.

I have always found it remarkable that a particular style and type of point would be standardized by Native Americans over such a large area for such an extended period of time, in this case from roughly AD 1250 or 1300 to around AD 1900, well into historic times. Why doesn't each little subgroup invent their own style? Why would they conform to a template that was adhered to

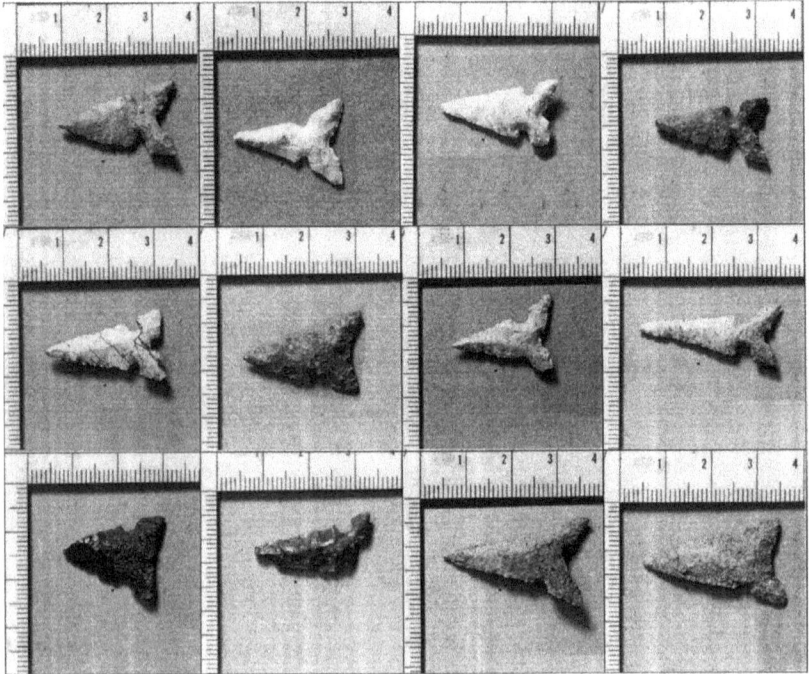

FIGURE 29. Desert Side-Notched points found in the northwestern end of the house.

FIGURE 30. Cottonwood Triangular points found in the northwestern end of the house.

over vast distances across language and cultural boundaries, and that survived for centuries? What is the mental construct that explains such behavior?

Because these DSNs are so standardized, and because they do reflect a rather specific time period (albeit a long one), they are thus referred to as diagnostic (as mentioned above); in other words, they represent something specific to the archaeologist, a time and place in the archaeological record.

The other point type found here was the Cottonwood Triangular, small points without notches. Their age and territory overlap with the DSNs but can actually be older, coming into common usage around AD 900, a few centuries prior to their later cousins the DSNs, and they also lasted well into the 19th century. But again, why the clear distinction? The points found here, 12 DSNs and 18 Cottonwoods, fall into one or the other category. The arrowhead maker had in his mind these two mental templates that he followed assiduously, making one type or the other, with no examples showing experimentation or deviation from the expected standards.

But while contemplating these anthropological questions of mental templates and standards over time and space, we can also simply stand back and view with appropriate awe the skill required to make such pieces. A single closeup photograph of a DSN (selected from the 30 arrowpoints found) reveals to us, the viewers, the intricate and complex pattern of flake scars that form the edges and the notches, the final product of a flint-knapper's unimaginably precise skill set.

The two artifact types, the arrow shaft straighteners and the points, led us to the interpretation that this end of the burned house was dedicated to the

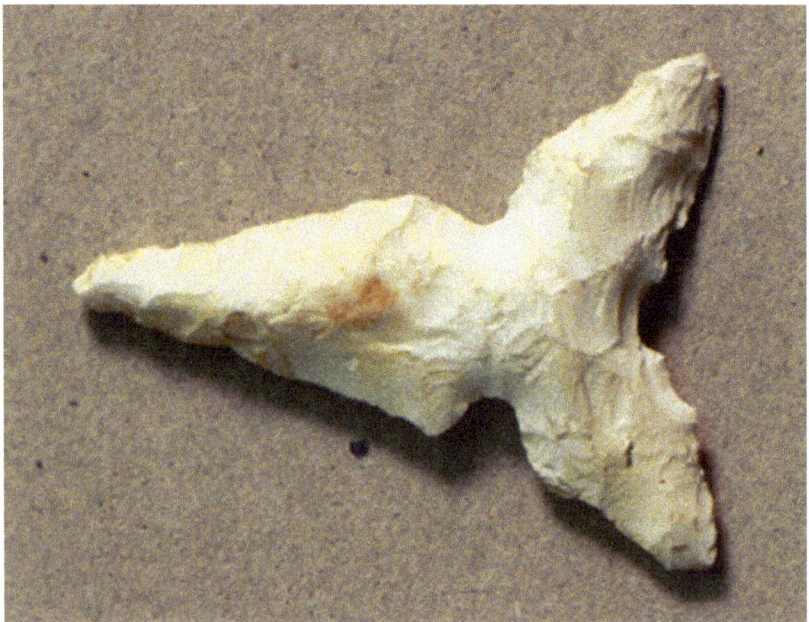

FIGURE 31. Desert Side-Notched point with detailed flaking scars.

arrow makers' craft, a sensational collection of 30 arrowpoints and 3 arrow shaft straighteners found together in the floor of a singular, extraordinary find, a one-of-its-kind for the Coachella Valley.

Religious relics: medicine man's tobacco pipes and tortoiseshell rattle. A picture began to emerge as details came to light. Thirty arrowheads and three arrow shaft straighteners in one place? These were not remains from a workshop; we found no telltale flakes or chipping debris (known by archaeologists as "debitage") that would have been ubiquitous if knapping had been going on there. These were intentionally placed, arranged for some purpose that we may or may not at some point fathom.

Then we found a clay tobacco pipe and remains of a tortoiseshell rattle, including the manzanita seeds that once rattled inside this sacred religious instrument. The owner of the 30 arrowheads was not only a master arrow maker, he was a medicine man and spiritual guide!

The clay pipe emerged gradually from the sand. Under careful brushing, it slowly became exposed, a mysterious double-rimmed reddish clay object, not immediately recognizable, but with a tapering stem on one end and a flaring conical bowl on the other. Further exposure with a soft-bristled paintbrush, powdery fine-grained ash and sand blowing this way and that with each brushstroke, the object finally appeared in full, a rough, hand-shaped clay pipe with a tapered stem, cone-shaped bowl, and a rare, widely flaring mid-girth protruding rim. Once out of the ground, it could be cleaned and photographed.

Near the pipe we found burned tortoiseshell fragments, many drilled through, mixed with charred manzanita seeds. We think that the manzanita seeds were once part of the rattle. The small burned little spheres were identified as mansanita during the subsequent laboratory analysis phase, just as the shell fragments were later identified as *Gopherus agassizii*, desert tortoise. The faunal experts tell us both the carapace and the plastron were present in the findings, the hard back shell and the under shell. If the pieces were not drilled, they might have been interpreted as food remains, but the holes in the shells tell us they were part of a manufactured implement and the ethnographic and archaeological literature have many examples of shell rattles drilled to let the sound out, the hard seeds inside the shell shaken in rhythmic, ancient song.

Cremated human remains. More than 1,500 pieces of cremated human bone were recovered from this feature that were positively determined to be human; another 4,000 pieces were probable. I must pause here to relate a very touching

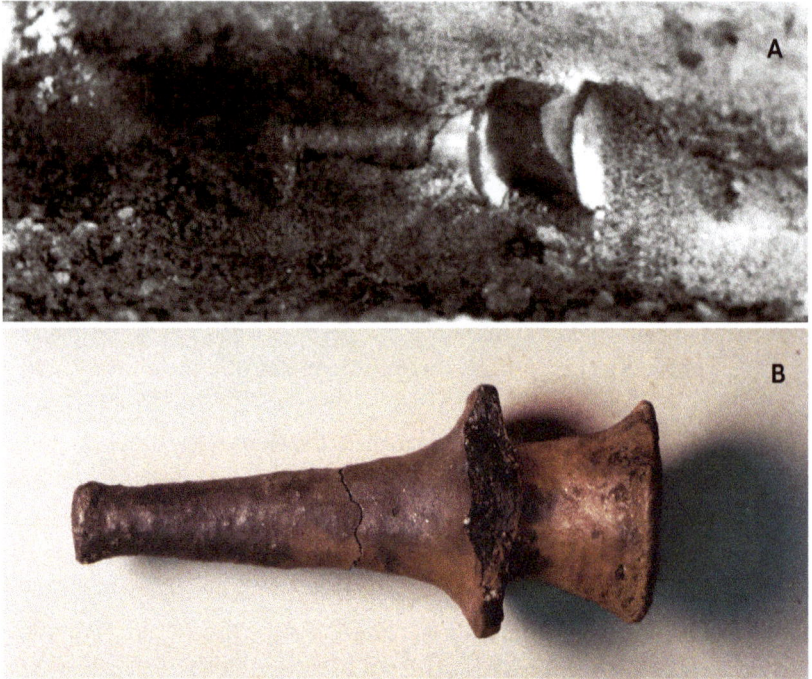

FIGURE 32. Clay pipe buried in the ash and sand (a) and later cleaned and reassembled (b).

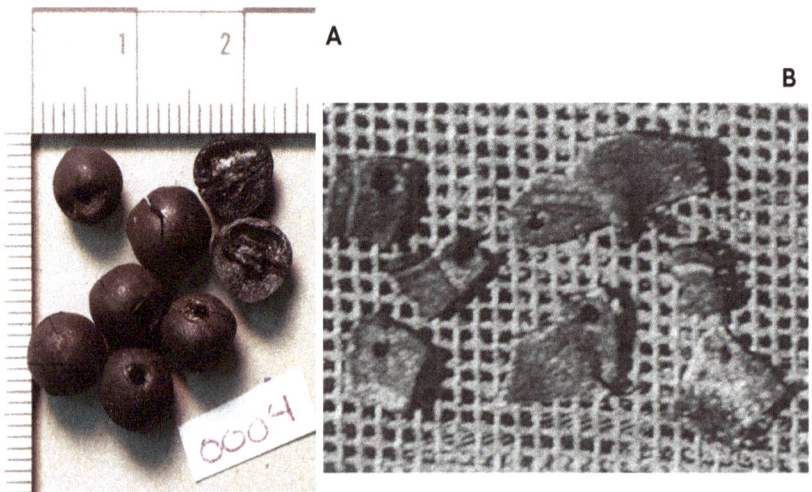

FIGURE 33. Burned manzanita seeds (a) and drilled pieces of tortoise shell (b), probably once part of a tortoiseshell rattle.

incident that happened to me recently (as of this writing). I was working on a repatriation issue, trying to get Native American remains returned and reburied, and I forwarded to a Paiute colleague of mine the details of the nature of the remains, how many adults, how many infants, how many male and female, and he reported back to me that I had caused him some considerable trauma, that he was a Middle East war veteran and suffered PTSD, and to hear of Native American ancestral remains talked about in such a cavalier manner caused him stress and sleeplessness.

I felt ashamed that I could once again be so clueless in dealing with my Indigenous friends—always learning hard lessons, continually it seems. But most seem to sense that I am at least trying, and they continue to give me access and friendship in most Native communities where I work.

The tiny charred and calcined (heated till they turn white) bone fragments clustered at the two ends of the feature and the pieces that could later be identified as to gender told us that there was a male and a female adult cremated in this house. The male remains were found mostly at the northwestern end where the arrowheads and arrow shaft straighteners were found. The female remains clustered around the southeast end, and sure enough, we found female-related artifacts there.

Before I go any further, I should tell how we dealt with not only the human remains but the feature in general. Any sensitive person, whether anthropologist or not, should be concerned when reading this about Native American consultation and participation. If I was reading something like this written by another archaeologist, I would be growing outraged that such a sacred site could be dug up and analyzed in such a callous manner.

We were not callous. First of all, we had tribal members working with us as archaeologists, Cahuilla crew members who were reporting back to the tribe at Torres Martinez exactly what we were finding. There were no secrets here. All information was freely shared with the tribe. Secondly, we specifically brought Torres Martinez elder and medicine man Ernest Morreo to the site for consultation.

Our geologist and archaeologist Harry Quinn, mentioned in previous chapters, went to Mr. Morreo's home on the reservation and picked him up and brought him to the site, where we showed him what we were finding and asked his counsel and advice. He blessed the site with prayer and tobacco smoke spread by eagle feather and gave us permission to continue our excavation

FIGURE 34. Torres Martinez medicine man Ernest Morreo (left) consulting with archaeologist/geologist Harry Quinn at the burned house site.

and analysis, as long as everything could be reburied when we were done. We agreed.

Mr. Morreo and others at Torres Martinez have said our work was of tremendous value because we were protecting ancestral remains from the developers. Others in the field, including other archaeologists and other Native Americans, have accused us of destroying sites by letting the developers continue to build over sacred ground. It's not an easy path to mark out and follow, this constant balancing act between the developers, city planners, archaeologists, and Indians: "not easy" is a gross understatement. I can only say in our defense that we always worked closely with Torres Martinez and that the elders there seemed to hold our work in high regard. When I sold my CRM company and moved out of Riverside County, they honored me with a luncheon and gifts (including a blanket) and a card. I will settle for that.

Crushed and blackened woman's artifacts. On the southeast end of the house was a large slab metate, a mano, a broken pestle, and remnants of at least two clay water jars (ollas), a food bowl, and a cooking vessel: food preparation implements arranged and displayed for the funeral pyre, a tribute to the eminent woman cherished and enshrined here in the southeast end of the house.

FIGURE 35. Slab metate, pestle fragments, mano, and pottery sherds at the southeast end of the house (see also fig. 27).

The slab metate had been intentionally broken into two halves before the burning, a common ritual practice to end the life of the piece, and one of the broken halves was broken again and one piece was shattered into fragments. Next to the metate were pieces of a long schist pestle, a two-handed implement for pounding and grinding in deep wooden mesquite-bole mortars.

A mano-sized piece of pumice appeared, highly abraded, a specialized tool perhaps with multiple functions. Pumice is a resource unique in Southern California to the Coachella Valley; it is a light, porous, hardened volcanic ash that floats on water, produced in not-so-ancient times in the volcanic fields of the southeastern Coachella Valley, blown by the shifting winds onto the shores of ancient Lake Cahuilla (see chapter 5), found today throughout the valley in archaeological contexts.

Four pottery vessels were reconstructed from the dozens of small pieces sifted and collected during the screening process, including an exquisitely shaped small water jar, once part of the woman's assemblage of household implements, prized possessions cremated with their matron.

Netting, mats, and cordage. Rare to the extreme, we recovered samples of netting, woven mats, and cordage from the burned feature. Two kinds of fiber were evident: agave and yucca. Two-ply cordage showed both clockwise and

FIGURE 36. Reconstructed water jar (olla) whose fragments were found under the slab metate.

counterclockwise twisting, and patterns of knots suggested netting. Cahuilla use of nets and cordage was essential and extensive in daily life: carrying, wrapping, hanging, fastening—an almost endless list of uses. It may well be that the people themselves were wrapped in netting for the cremation fire, a practice documented in the ethnographic sources. Or they may have been wrapped in woven mats, mats that once covered the floor and provided sleeping beds. Rare indeed were the remains of these woven and twisted artifacts recovered and recorded by the archaeologists.

Shell beads. Shell beads from the coast represent wealth and prestige, and the cremation remains found here were no exception. One hundred sixty beads were recovered in the screening, all but one being the Olivella cupped variety, made from the thickest part of the wall of the Olivella shell. These bead types are manufactured along the Pacific Coast, in particular among the Chumash of Ventura and Santa Barbara Counties, and were prized trade items reaching Arizona and points east. The famous Mohave traders came from the Colorado River along desert trails, sustaining themselves on chia seeds carried in small skin pouches, running hundreds of miles, their bodies painted for power and protection, often running all night in the summer months, to trade for the highly prized shell beads on the coast.

FIGURE 37.
Knotted cords,
probably remains
of netting.

FIGURE 38. Double-twined yucca fiber cordage.

FIGURE 39. Fiber matting.

FIGURE 40. Olivella cupped wall beads.

Burned remains of a limpet shell were also found, suggesting that the Olivella shells and the limpet shell were strung together in a power necklace.

Age. The single radiocarbon date from the burned house was AD 1730, plus or minus 50 years, just before the so-called historic period, on the very cusp of the non-Native, or Spanish, invasion. The first Spaniards to permanently occupy California arrived from Baja California in 1768, mostly keeping to the coast, and eventually setting up their mission system. Somewhat later the first non-Natives, again the Spanish invaders, made their way inland in the later 1700s.

But tragically, the invasion by the Spaniards (California was part of New Spain at that time) may have been preceded by their Spanish-born diseases, especially smallpox, which was spreading rapidly across North America by the 1600s. It is only speculation, but perhaps the twin cremation of man and woman at what today in known as Indian Wells Country Club East Course, was the result of death by smallpox. For a couple to die together and to be cremated together suggests unusual circumstances, but again, the cause of death could not be determined by the scientific analysis of the remains.

Arrowweed and wood frame. Analysis of burned plant remains found an abundance of arrowweed, a hard woody shoot found throughout the salt flats of the Coachella Valley that was used to weave the Cahuilla houses, passed in and out through the heavier wood framework, a long-lasting building material. (As the name implies, these stalks were also used to make arrow shafts.)

Larger pieces of charred wood lay among the artifacts and remains (see fig. 25), probably once part of the overall structure, the *kish*, or Cahuilla house.

Final analysis. My intention here is to honor the man and woman that were so venerated in their own time, that were so respected that upon their death their family, fellow villagers, and probably emissaries from distant lands gathered and certainly held ceremony. I often reflect just how miniscule is the picture we get from the archaeological record, usually only the pottery and stone (chipped or ground) and very little else. But the ceremony, the gatherings, the songs, dance, music, performance, laughter, and weeping, leave not a trace.

We get an idea what it was like from contemporary practice in more traditional communities, like Torres Martinez, where funeral ceremonies go all night and into the next morning, where the wake songs are sung and cloth is hung to protect the space. When Biff Andreas passed (as told in chapter 2), they erected a special "birdhouse" for the visiting singers, who performed all night, while in the main hall the *noquanish* songs were sung by singers from

Pechanga, one of the very few groups at that time who knew those songs. At dawn the body was carried out to the funeral pyre and as the flames grew higher, the visiting singers tossed their rattles into the fire, the ultimate tribute.

Throughout Indian land still today, the deceased are buried or cremated with their personal belongings. The burned house we blessedly uncovered at Indian Wells Country Club East Course contained the remains of a man and woman and their sacred possessions, apparently arranged in a funerary tableau that was never meant to be disturbed! We archaeologists must remember, when human remains and their possessions appear at construction sites, it is the ultimate desecration. Remains that were interred with sacred rites and ritual should never be disturbed, but they are. Everywhere. The best we can do is treat them respectfully, work closely with the nearest tribes, and get them reinterred if that is the tribe's wishes.

In this case, conferring with tribal medicine man Ernest Morreo, we collected the remains, analyzed them, wrote them up into a report, and then performed a reburial following Mr. Morreo's instructions. Tribal intentions everywhere are that remains go back in the ground as close as possible to where they were found, but when that's not practical, in an alternate location designated by the tribal elders.

Here in Indian Wells, the golf course developer helped us choose a location very close to the original find, a place designated on the planning maps as an easement with restricted access and likely to not have future development. Using a backhoe, we dug a very deep pit, deeper than would likely be disturbed in the future, and with Mr. Morreo and other tribal members present, we slowly lowered the collected remains back into the ground, where we hope they will be undisturbed in perpetuity.

We honor the man and woman cremated there some three centuries ago. It was a marvelous discovery bordering on miraculous, and their memory will now live on in print thanks to this book. May Cahuilla and non-Cahuilla alike pay their respect.

CHAPTER 4

Bighorn Sheep Management

B IGHORN SHEEP ARE, and always were, hugely consequential in Cahuilla life. They are part of the creation story; there exists an entire body of Cahuilla bird songs, sung by traditionalists, known as Bighorn Sheep Songs; and the winter sky is dominated by the Bighorn Sheep constellation, known to most of us as Orion. They are the guardian spirits over deer and antelope.

Fields of cairns. Archaeologists have now found evidence, overwhelming evidence, spread throughout almost every cove where the steep San Jacinto Mountains dip to meet the lower desert floor, from Palm Springs to La Quinta (and Anza-Borrego Desert State Park and elsewhere), telltale remnants of specialized rock cairns: small piles of granitic cobbles, that once formed sophisticated patterns on the sides of canyons, laid out in rows and in clusters, that were part of Cahuilla bighorn sheep management strategies in preinvasion times.

These cairns are important for several reasons, not the least of which is that they are unknown outside of the specialized archaeological community. How many of you readers subscribe to the *Journal of California and Great Basin Anthropology*? An important article about these bighorn sheep–related features by archaeologist Joan Schneider appeared in the *JCGBA* in 2014, and multiple other reports are buried in the aforementioned Eastern Information

Center file cabinets. But they are unknown to the general population of the Coachella Valley (or any other part of California), and therefore deserve a chapter in this book so that at least somebody will have heard of them. Some 40 years ago now, archaeologists from UCR first noted these cairns in a canyon near Palm Springs and proposed what they were for. But could it be proven?

By the time I personally began doing archaeology in the Coachella Valley (early 1990s), several groups of these cairns had been found, identified, and published in academic journals. Then a La Quinta water tank project in 1995 gave me a cluster of some 220 cairns for my very own research, unexpected and unforeseen. A rather typical CRM job, a phase I survey of 80 acres, turned out to be a treasure trove of priceless findings, and an exceptional opportunity to combine contract archaeology with scholarly investigation.

An aside on CRM archaeology. There's a strange dichotomy in the archaeological world, at least there used to be, between CRM archaeology and so-called scholarly archaeology, with a class distinction of ranked status: academics disparage contract archaeologists, and contract archaeologists hope and dream of someday moving up (a perceived upward advancement) into academia. I had an employee once who said she hated doing CRM, it was selling out, working for the developers (like land developers were the enemies). Occasionally, I would give a speech to my employees to try to convince them to be proud of doing CRM. After all, we were on the front lines protecting sites, and we should work to find common ground with developers and government agencies and tribes, difficult to do but doable (usually).

Another aside: A friend of mine was a surveyor for Los Angeles Department of Water and Power working in the Owens Valley, and, along with his coworkers, played a trick on archaeologists who were surveying a project route. The LADWP surveyors surreptitiously made some rock piles along the job route and watched with glee as the archaeologists came along and recorded them as historical features. Now if the archaeologists were more experienced, or at least had a more experienced supervisor, they would have detected that these fake cairns were recent. But alas, they were fooled (at least as my surveyor friend tells it).

Yet another aside: A fascinating, to me, aspect of contract archaeology is how the archaeologist has no control over where and when he or she surveys and digs. Academic archaeologists plan their work, develop research designs, submit grant proposals, do surveys and digs, and write up and publish the results in academic journals. CRM archaeologists get phone calls or emails from prospective clients, submit cost proposals, do jobs, write up their results, and usually do not publish their findings. The locations of CRM job sites are pure serendipity (luck?) driven by the housing market, highway rights-of-way, power line routes, or, as in this case, a new water tank for Coachella Valley Water District.

The water tank cairns. When we did the survey for the new water tanks on the slopes of the foothills behind La Quinta, we discovered more than 200 small cairns that—judging from their looks—were old, not recent. In fact, cairns, or small rock piles, are very difficult to judge regarding age and function, but the ones at the La Quinta water tank site did look old, the way the

FIGURE 41. Approximate location of field of cairns northwest of present-day Lake Cahuilla Veterans Regional Park.

FIGURE 42. A typical cairn; the scale is about six inches long.

individual rocks had settled down into themselves with a buildup of wind-blown sand and vegetation growth in and around them.

The 200-plus cairns that we discovered at the La Quinta water tank site were rather consistent in size: a typical cairn consisted of about a dozen rocks ranging from 6 to 16 inches in size, piled about 12 inches high with a base about 28 inches wide. The cairns were spaced out over an area of sloping, uneven terrain some 2,000 feet long by 250 feet wide.

What an opportunity! A literature review gave us multiple examples of fields of cairns in the coves on the western edge of the Coachella Valley, and these same reports, gathered by the records search we did at the Eastern Information Center (see chapter 1), presented some hypotheses about their function. Now, thanks to a completely random contract job, unplanned and unforeseen, we had a chance to study a cairn field in depth and to test competing hypotheses, all paid for by the Coachella Valley Water District as part of our proposed mitigation plan to lessen the impacts of a construction project on an important archaeological resource. Sometimes the system works!

Competing hypotheses. Based on our literature review, there were three distinct hypotheses that could explain cairn fields like the one at the La Quinta water tank site and other similar fields in the coves of the western Coachella Valley, and we had the chance to test each hypothesis and thus come closer to nailing

FIGURE 43. A typical section of the cairn field, showing layout and distribution of some 15 of the total of 220.

the correct explanation: (1) game management/hunting; (2) caches for food storage or other items; (3) unknown cultural practice, religious or otherwise.

Data collection. "There's nothing like data to ruin a good theory," the old saying goes, so before expounding one's theory on this or that, it's a good strategy to collect lots of data. In fact, getting raw data should always be the first step in any scientific inquiry, and if the data can be arranged in categories and if the subject of the study can be thus catalogued, all the better. With our cairn fields, we set out to record them all, each one, as accurately and thoroughly as possible, but without unnecessarily destroying the cairns that could be saved (you've probably heard about the bristlecone pine tree, perhaps the oldest living plant in the world, that was killed when a scientist who was measuring the tree got his boring tool stuck in the trunk and a nearby helpful ranger cut the tree down to retrieve the tool).

Archaeology is destructive; you can never put back what you have excavated, and we soon learned that we could not accurately count or measure the rocks in each cairn without disassembling the cairn and thus destroying it, so we determined which cairns were destined to be destroyed by the water tank project and which ones lay outside the project construction zone, and we reserved our destructive testing for the ones that would be destroyed anyway,

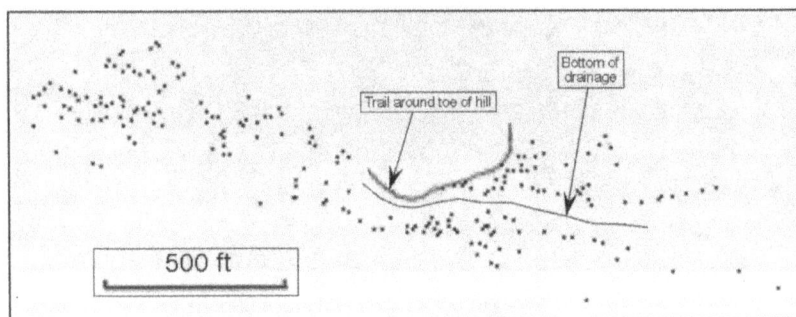

FIGURE 44. Pinpoint mapping, thanks to Coachella Valley Water District, gave us a permanent record of where each cairn was.

and put into practice our nondestructive data collection for the ones that could be preserved. The cairns that lay outside the construction zone were left intact. We estimated the numbers of rocks and their measurements without taking the cairns apart so that they could be preserved, hopefully in perpetuity.

The cairns themselves were mapped in place without disturbing them, and to that end, the Coachella Valley Water District provided a survey crew that pinpointed each cairn for us (this was before GPS), creating an accurate and complete map. Not surprisingly, an additional half dozen cairns were discovered during the mapping process.

Counting the rocks. I mentioned that we accurately counted the rocks in the cairns that were destined to be destroyed by taking them apart and measuring and counting them, but to avoid unnecessarily destroying cairns that could be preserved, we estimated the counts and measurements for those cairns without disassembling them, which led us to wonder how accurate our estimates were; so we did a simple test. We selected 15 cairns that were destined to be destroyed and estimated the counts and sizes of the without disturbing them, and then we took them apart and did real counts and measurements. We were surprised how much we underestimated the number of rocks in each cairn, by an average of 39 percent! Such a discrepancy did not affect the hypotheses we were testing, but it did teach us a lesson about the accuracy of doing estimates vs. real counting.

Desert varnish. We hoped that desert varnish might help us understand at least some aspect of the cairn field phenomenon. Desert varnish develops on rocks over centuries of direct exposure to sunlight. As we disassembled the cairns, we noted the presence or absence of varnish and whether it occurred on the upper side or underside of each rock, or both. "Both" turned out to be an important clue. Typically, a rock that lies undisturbed and half-buried in the desert floor develops varnish on its exposed surface. It also develops a whitish band known as a "ground-line band" at the interface between it and the ground matrix surrounding it.

We found rocks in the cairns that had varnish on one side and the telltale band around the circumference, the result of someone pulling a rock from the desert floor and using it to build a cairn. But we also found rocks with varnish on both sides, and rocks that had ground-line bands that did not align with the ground and that were developing new ground-line bands that did align with the ground. In other words, a rock was taken from the desert floor and set in the cairn with the varnish side down. Then, over time, varnish began to appear on the upper newly exposed surface, and rocks at the base of the cairn began to develop new ground lines.

Finding rocks with desert varnish on both sides and more than one ground line told us that the cairns were of very great antiquity. They were built so long ago that varnish and ground lines developed after they were built. There is no way (yet known) to put a finite date on these processes, but at least multiple centuries (if not millenia) have passed since the cairns were built.

Excavations. To test hypothesis 2, which was that the cairns might have been caches for food storage or other items, was easy; we simply excavated beneath a sample of them to look for any evidence of items having been stored there. After removing one rock at a time, a one-meter-square grid was cleared down to bedrock with all material screened through ⅛-inch hardware cloth. We found no evidence of food remains: no husks, seeds, pods, branches, or any other indication of vegetation; neither bones, teeth, or any other sign of animal remains.

Secondly, a review of the ethnobotanical literature finds no examples of food caching in rock cairns by the Native peoples of the Coachella Valley. The plant foods used by the Cahuilla people—most importantly mesquite—included numerous varieties of seeds, blossoms, leaves, roots, stems, etc., but the literature makes no mention of storage in rock cairns. The dominant plant species on site today, the creosote bush, was used mainly for medicine, and certainly was not stored in quantities requiring rock structures for protection or preservation.

Ethnography. It has just been mentioned that none of the ethnographic literature mentions rock cairns for food storage (or any other purpose), but what

FIGURE 45. One-meter excavation units under a sample of the cairns confirmed they were not used as caches for storage.

about doing modern ethnography? There were some very traditional Cahuilla elders living at Torres Martinez, and I thought this was an excellent opportunity to interview at least one of them, if he would oblige me.

Saturnino "Sat" Toro lived with his wife at "the Hill," the local nickname for the Cahuilla housing tract out by the dike at the west end of 64th Avenue, a location near the historic village of Toro and the old Toro Cemetery. Sat and his wife liked to sit in their chairs under the solid shade of the fruitless mulberry tree in their front yard when hot summer days slowly turned to evening. I approached the couple timidly one late afternoon and found Mr. Toro gracious and responsive. Did he have any idea what the cairns may have been for? No, he said, no idea.

But what he did tell me about bighorn sheep hunting was crucial. He recalled that he and his brothers hunted bighorn sheep when he was young (using rifles of course, not bow and arrow). He said they would find a place to wait for the sheep, a place where they knew the sheep would come. When the sheep approached, they had to be extremely still; the slightest movement or disturbance would scare the sheep and the hunters would not get a second chance. One shot! This hunting technique strongly supports our working hypothesis that the cairns worked as a passive hunting complex that guided the sheep to where the hunters were waiting.

But before going on, I need to tell the funny part of my interview. I had noticed on previous visits to "the Hill" that there were small cairns across the front of Mr. Toro's yard, between the street and his front lawn, and I recalled these when we started on the La Quinta Water Tank project.

Hypothesis 3 in our list of hypotheses was that the cairns represented some kind of cultural or ceremonial practice, religious or otherwise. Was there some special significance to a line of small rock piles that perhaps Mr. Toro could tell me about? I asked him, hoping for some insight into ancient Cahuilla sacred traditions perhaps still practiced by one of the most highly respected elders on the Torres Martinez reservation. What were his cairns for, I asked, the ones in a line separating his front yard from the paved street beyond? He looked at me solemnly and explained: "I put those there to keep drunk drivers from driving up on my lawn." Wow!

The hunting hypothesis. Comparing the cairns at the La Quinta Water Tank project with other previously recorded fields of cairns in the Coachella Valley, we found further support for the passive hunting hypothesis. The cairns probably held brush or branches upright, creating a visual barrier that would have directed the sheep to where the hunters were waiting.

All the groups of cairns thus far recorded in the Coachella Valley were constructed along edges of small, secondary drainages that flow out of side canyons along the bases of mountains. Bighorn sheep prefer terrain that is rocky and broken by washes and that affords an opportunity to escape up steep sides of nearby mountains. Each cairn site had a larger, wider drainage nearby, but use of that drainage by the sheep would have exposed them to open terrain, a feature that bighorn sheep instinctively avoid.

One day while working at the site, we got a chance to observe bighorn sheep behavior ourselves. We saw a small herd of eight bighorns some 100 yards upslope from us in a small drainage. When one of us pointed at them, they instantly fled straight up the escarpment next to the drainage. Most Native hunting complexes that have been recorded archaeologically in the western states are remains from drives and corrals, where animals such as pronghorn or rabbits are driven into confined areas and then dispatched. Bighorn sheep cannot be driven.

The absence of artifacts in the Coachella cairns sites is consistent with the passive hunting hypothesis, directing animals' movements rather than driving them into corrals. If the sheep were hunted and killed in the cairn fields, we would have found evidence of such in the form of arrowheads or stone chips from butchering tools, but we did not. The sheep were guided by brush walls to where the hunters were waiting, probably at the bottom near the water holes.

Brush walls. Replicative archaeology is extremely important in our field. If you want to know how mortar holes in granite boulders were used, try making some yourself. Is the hole a natural result of pounding seeds and nuts, or were they made to be deep in the first place and then used for food preparation? One can speculate about questions such as these, but by replicating such behavior, you learn so much more. Stringing a bow, shooting an arrow, making an arrowhead—these are all activities that provide answers to research questions that reading and thinking can never achieve. With this in mind, we built some brush walls with materials at hand.

At a nearby wash we found loose rocks and dead branches from palo verde and smoke trees, and we proceeded to create rock mounds that supported branches upright. Success! Gathering rocks at random, piling them around the bases of large dead branches that spread out above the cairns, spacing them so that the ends of the branches just touched, the modern cairns that we created looked almost exactly like the ones at the site. They had a similar number of rocks stacked to a similar height and were spaced apart from each other just like the ancient cairns from untold centuries earlier.

A satisfying conclusion. Taken together, all lines of inquiry supported the passive hunting hypothesis, which had been proposed by previous archaeologists but perhaps had not been so thoroughly investigated. The detailed data collection, the extensive literature search, the ethnographic inquiry, and the replicative studies all pointed to the same conclusion: that the Cahuilla people of the Coachella Valley practiced a highly skilled and complex system of game management and food procurement at a time no longer in memory or in recorded history or even in oral history.

An archaeological study, not thought up in the halls of academia, but required by state and local ordinances in advance of a major construction project, produced extremely important findings that supported and corroborated what other practicing archaeologists had found, studies that up to now had not been told to the interested public. Practically every cove along the west side of the valley has, or did have, remnants of these once all-important bighorn sheep walls, brush barriers stood up by small piles of rocks to direct the sacred animals to their destiny, an important part of Cahuilla culture that has been lost in practice but retrieved by archaeological inquiry, and now published for the first time in the popular press.

FIGURE 46. Modern replicative study produces cairns almost exactly like the archaeological ones.

FIGURE 47. Randomly collected rocks piled to create a brush wall of dead palo verde and smoke tree branches result in spacing very similar to the spacing of the cairns at the archaeological site.

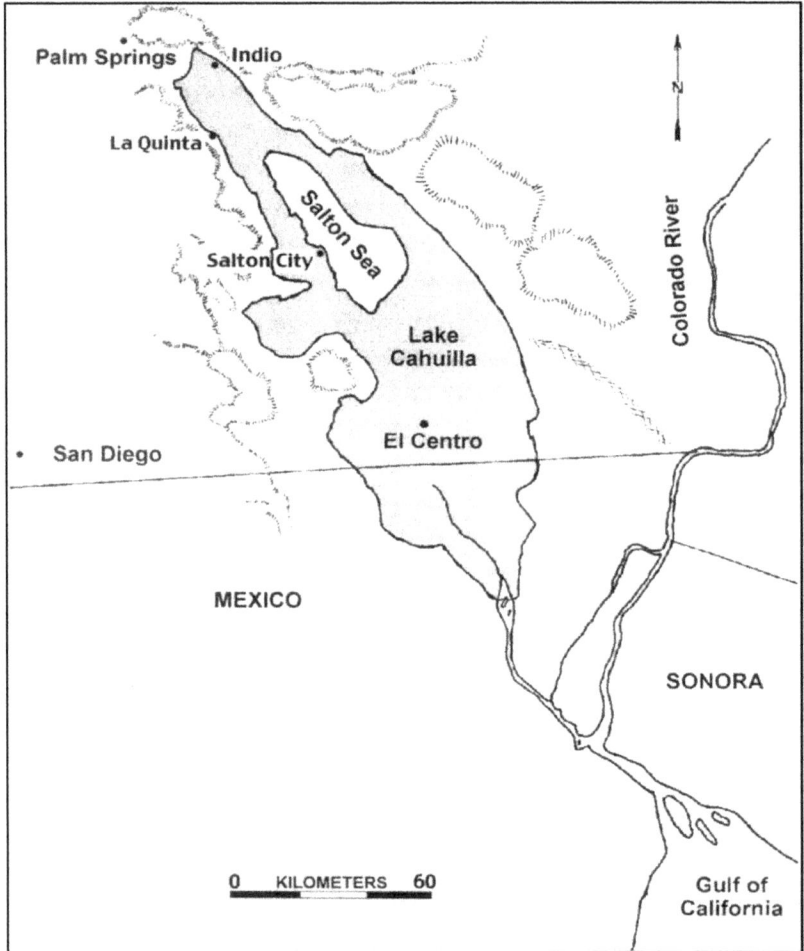

FIGURE 48. Comparing the size of Lake Cahuilla with present-day Salton Sea.
(Kenneth Gobalet and Thomas Wake, "Archaeological and Paleontological
Fish Remains from the Salton Sea, Southern California," *The Southwestern
Naturalist* 45, no. 4 [2000]: 515.)

CHAPTER 5

Ancient Lake Cahuilla

NO SINGLE GEOGRAPHICAL phenomenon had greater impact on Native lifeways in the Coachella Valley than the repeated infillings and desiccations of ancient Lake Cahuilla—from a bone-dry sink of salt and soda to a massive freshwater lake and back again, at least seven times in the last 2,000 years.

Today's Salton Sea is but a poor and polluted puddle, one-sixth the size of ancient Lake Cahuilla, that resulted from an engineering accident on the Colorado River that sent river water inland instead of to the Gulf of Mexico. It took two years, 1905 to 1907, to stop the flow of water and redirect it back into controlled channels, but before they fixed the breach, the water had created what is now the Salton Sea.

If only they had allowed the so-called Salton Sea to dry up the way all the previous Lake Cahuillas had dried up, we would today have a dry lake bed and thousands of acres of irrigable agricultural land, instead of a super-saline, contaminated, odorous sump plagued by massive bird and fish die-offs. Agricultural runoff from the surrounding farmlands, steeped in pesticides, drain into the sink and keep the Salton Sea from drying up.

But the Salton Sea is not the subject of this chapter. Instead, we are talking about the many incarnations of freshwater ancient Lake Cahuilla and how the Cahuilla people adapted and strategized and used the lake and made life-affirming choices along its banks, and how modern archaeology plays a role in understanding these ancient lifeways.

Lake Cahuilla formed naturally and periodically when the Colorado River delta built up so much sand and silt that the river began to seek alternative routes, which, when pushed by overly wet winters or perhaps extremely rapid snowmelt in the Rocky Mountains, burst through its banks and flowed westward and northward, filling the Salton Trough. After some years (nobody knows how long), the river would regain its old ways and empty once again into the Gulf of California. The lake would then immediately begin to evaporate, perhaps drying up completely within 30 to 50 years.

Dr. Phil Wilke at UCR was the true pioneer in these studies. His doctoral dissertation (1976) and later articles combined ethnohistorical research with coprolite studies and radiocarbon dates from stratified cultural deposits to signal the frequency and duration of past infillings of the Salton Trough. Multidisciplinary research continues to this day.

Ethnohistory. I have just introduced two terms, "ethnohistory" and "coprolite studies," that may need explaining. Ethnohistorians use historical documents to piece together past cultures, descriptions of people and their beliefs and behaviors, the stuff of anthropology but derived from data and documents that originally were just recording history. These include, for example, travelers' diaries, census records, administrative reports, personal letters—material in which the original authors were not necessarily trying to describe people and their culture, but from which such information can be gleaned.

Wilke researched early travelers who came across the lower Colorado Desert who reported their observations in the 16th through 19th centuries and found no mention of a large standing body of water with Indian villages around its shoreline. The Díaz expedition of 1540 saw no evidence of a lake, neither did the Oñate expedition of 1604–5, nor did Father Kino's travels in 1701 and 1702, and in the 1770s at least three Spanish expeditions crossed the region without reporting the presence of a lake.

Wilke suggested that if there was a late high stand of Lake Cahuilla, as some radiocarbon dates suggested, the lake must have come and gone between traveler X and traveler Y, when there was a period with no traveler's accounts. It turns out that one of our own digs in the City of La Quinta, right at the elevation where the highest lakeshore would have been, found a fish-roasting pit with charcoal and thousands of fish bones indicating a Native presence at a high shoreline.

When UCLA analyzed the fish bones, they found that two species of fresh-water fish dominated the collection (razorback sucker and bonytail chub) and that those two species today still predominate in the Colorado River, and the date we got from the charcoal fit neatly into Wilke's scheme, a late high stand around the years 1620–30 that no non-Natives saw.

Remarkably, more recent work by Thomas Rockwell at San Diego State University has found yet another even more recent high stand, based on radio-carbon dates from dead tree stumps, and once again, it fits smack in between two dates when travelers did not report seeing a lake, in this case between 1702 and 1770. The new date for a high stand supplied by Rockwell is 1726 (plus or minus 7 years). This means that the lake filled and then evaporated in less than 70 years. How dramatic this must have been for the Cahuilla people!

Coprolites. Returning to the other of Wilke's specialties you may not have heard of, what are coprolites, or even better, human coprolites? Dried and fossilized poop—yep, the study of ancient human feces from the desert sands of the Coach-ella Valley. In one report, Wilke had collected 1,000 specimens in a concentrated area of only four yards across, answering a question I had in fact never heard asked in all my years in archaeology: did the local Native Americans have latrines, or did they just poop anywhere? This profoundly consequential research question appears to have been answered, there in Myoma Dunes, near Indio, where 1,000 coprolites were retrieved from one small sand dune by Wilke and students in the 1980s.

I remember when I first arrived at UCR in 1990 to become director of the Archaeological Research Unit, the place was in a tizzy because a widely read and highly respected columnist for the *Riverside Press-Enterprise* had just pub-lished a column making fun of UCR's archaeologists for studying fossilized poop, especially, I recall, the details on how to rehydrate the "matter" to break it down and reveal its secrets. A PhD graduate student was composing a rebut-tal, a letter to the editor, upset that the public might find his work amusing, worthy of a bit of friendly ridicule. A graduate student with no sense of humor is not hard to find. How can you not smile at the thought of scientists rehy-drating feces to study its contents?

In fact, coprolite studies did provide a remarkably detailed picture of preinva-sion life along the north shore of ancient Lake Cahuilla. Native diet and resource procurement were reconstructed, telling us what the Cahuilla were living on at that time: marsh resources as well as dry land, aquatic birds and fish as well as rabbit and deer. Even pollen is preserved, which can tell us about seasonality of occupation. Were the sites temporary camps or year-round villages? Answer: both types were present. Thank you, Phil, for your conscientious, finely detailed, and thorough analysis of human "remains" (and forgive the smiles).

The shoreline. The highest shoreline that Lake Cahuilla could achieve was 13 meters, or 42 feet, above sea level. At that height the water began to flow out the southern end of the Salton Trough into the Gulf of California. Scientists still debate how long the lake stood at its high-water mark before beginning to shrink, but the stark clarity of its old shoreline on the cliffs south of La Quinta suggests it stayed at that level for quite some time. It is dramatically visible on the rocky slopes from La Quinta on down the base of the Santa Rosa Mountains to Travertine Point and beyond.

Apparently, it reached the same high-water mark repeatedly, because scientists have found multiple layers of tufa, a rough coral-looking deposit of calcium carbonate that coats the submerged rocks. Tufa deposits up to two feet thick have been found on some boulders, and slicing into them can reveal stratigraphic layering from repeated infillings of the lake, although attempts to date the layers have so far proved unsuccessful.

FIGURE 49. The 42-foot high lake stand appears on the cliff faces south of La Quinta.

FIGURE 50. Example of tufa buildup on rocks near Travertine Point.

While the 42-feet-above-sea-level line is clearly visible on the cliffs, it can also be traced across the flats of La Quinta and points east, not by tufa deposits but by contour lines on topographic maps. As it turns out, the high stand of ancient Lake Cahuilla passes right through the City of La Quinta. I found a 1959 topo map that has the 40-foot contour clearly marked, and we can see that it passes right through the intersection of Avenue 50 and Washington Street.

That's why the archaeology of La Quinta is so extremely rich. There were Cahuilla villages and campsites everywhere along the old shoreline. I mentioned earlier the fish-roasting pit that we found that had the Colorado River fish and charcoal that allowed us to confirm the date of a recent high stand. That site was right on the north side of Avenue 50 a little west of Jefferson Street. A whole cluster of sites was found on the southwest corner of Avenue 50 and Jefferson Street. And so on.

Great credit must be given to the City of La Quinta and their citizen oversight, in particular a Ms. Barbara Irwin, head of the cultural/historical committee that insisted that archaeological studies be done as part of the explosive development during the 1990s. What appeared to be barren, desolate sand

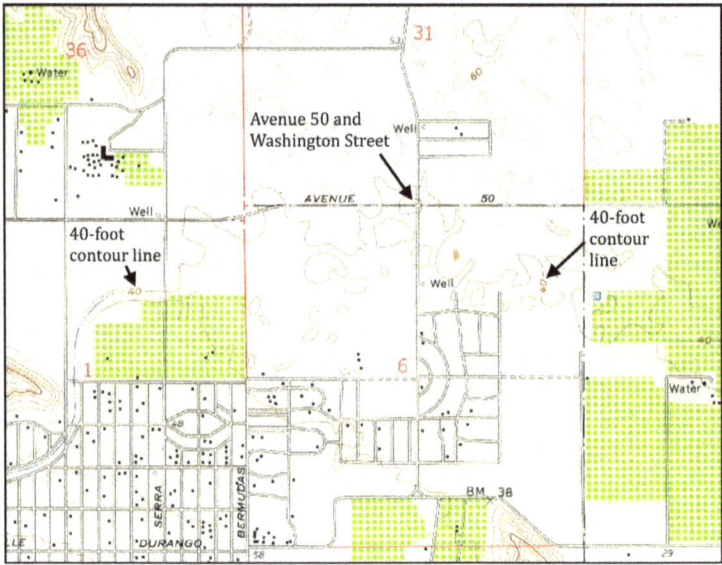

FIGURE 51. The 40-foot contour line passes through La Quinta.

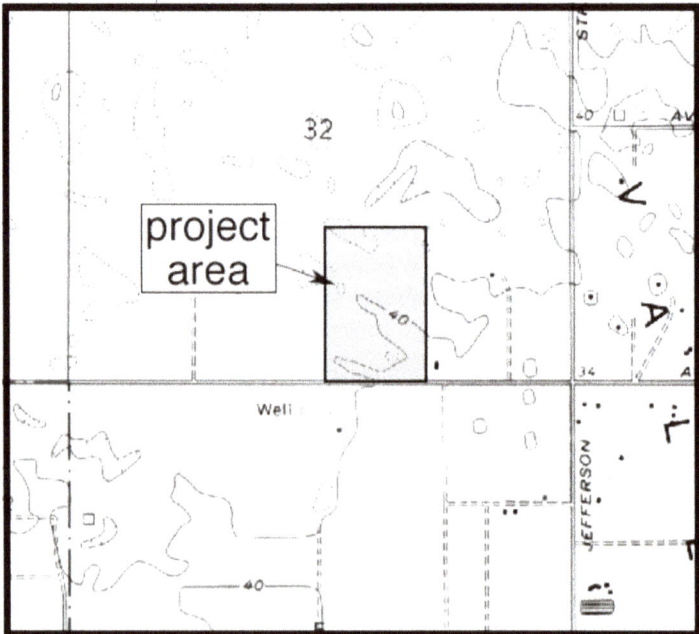

FIGURE 52. The spot near present-day Aveunue 50 and Jefferson Street (note the 40-foot contour line) where an Indian fish-roasting pit was excavated, proving a high lake stand in the 17th century.

dunes could have easily been written off by town councils or planning departments eager to issue grading and building permits, but La Quinta, probably
more than any other city in the valley, required archaeological surveys, testing,
and mitigation in front of almost all building projects, resulting in discoveries
and findings of great import, for example, pushing back the antiquity of Native
settlements by 2,000 years, as described in chapter 2.

Fish traps. As the lakes came and went, the Cahuilla people took advan-
tage, exploiting the waters and wetlands for life-giving plants and animals, fish
above all else, marvelous sources of protein. Where the shoreline lay across
boulder fields along the base of the mountains, such as near Avenue 66 and
Jackson Street, the people made large five-to-eight-feet-across bowls in the
boulders right at the shoreline. Wind-borne fluctuations in water level and
waves lapped over the rims of the bowls and brought fish, or fish would seek
out such sheltered areas to deposit eggs. The waiting Indian fishers then netted
or speared them, taking them to their home villages for roasting, leaving traces
for archaeologists to find centuries later.

FIGURE 53. Stone fish traps near Avenue 66 and Jackson Street. (Lindsay
Porras, "Environmental Diversity and Resource Use in the Salton Basin of
the Colorado Desert," master's thesis, California State University at San
Bernardino.)

Two principal kinds of fish traps are known today: the bowls in the boulder fields just described, and V-shaped (or U-shaped or J-shaped) rock alignments on the more gently sloping sands and gravels away from the base of the mountains. On the more level areas, rocks were laid out with the narrow opening facing downslope into the water with the legs of the V pointing upslope. It's not clear to me, from reading the literature, if the fish swam around into the big end and then were trapped trying to escape down through the narrow end, or if they came in the narrow end.

One needs to read the ethnographic literature where similar fish traps are still in use in other parts of the world, or where eyewitness accounts recorded observations of them, to determine how they worked. Coachella Valley fish trap research is crying out for replicative studies (mentioned in chapter 4). I know of no such studies to date, but I've heard lots of opinions on how the fish traps worked. Some have even suggested Lake Cahuilla was large enough to have tides, and that the fish traps were like tide pools where the fish could be easily caught.

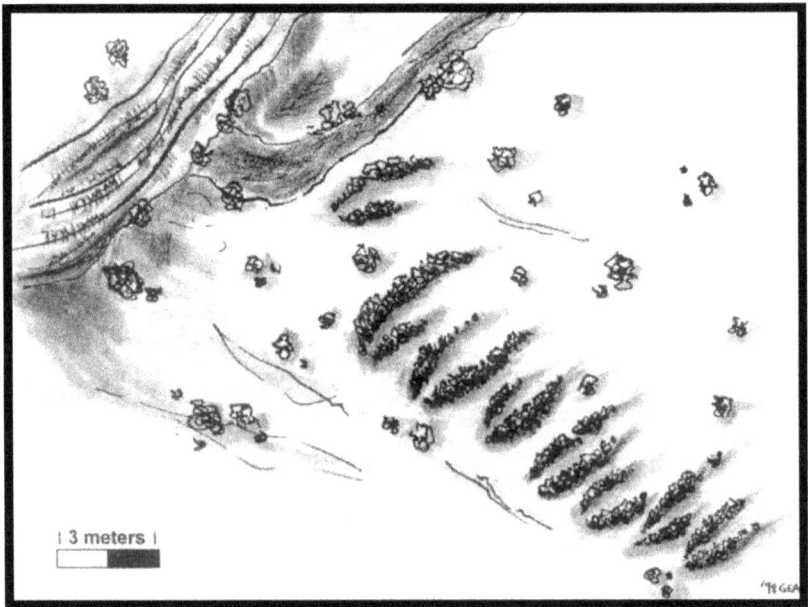

FIGURE 54. Sketch of V-shaped fish traps. (Kenneth Gobalet and Thomas Wake, "Archaeological and Paleontological Fish Remains from the Salton Sea, Southern California," *Southwestern Naturalist* 45, no. 4 [2000]: 514–20.)

One salient feature of all the clusters and rows of fish traps—and they have been found all around the old shorelines—is how they mark the shore-line retreat, the drying up of the lake. Rows of fish traps were built along the shoreline, and then as the lake began to evaporate, more rows were built along the new and lower shorelines, resulting in lines and lines of fish traps, each marking a subsequent lower elevation of the water, a fascinating his-torical timeline made visible for generations to come in the rocks and sands of the desert floor.

Obsidian Butte. At the south end of the present-day Salton Sea lies a series of obsidian extrusions, places where black liquid glass squeezed out of the earth and solidified, providing Native peoples excellent material to make sharper-than-razor knives, arrowpoints, and cutting tools of many kinds. An obsidian knife edge is 10 times sharper than the sharpest steel razor blade, but obsidian is brittle and tools made from it break easily.

In recent years, archaeologists have sourced obsidian by analyzing its min-eral content to create obsidian signatures of volcanic sources that can be com-pared with flakes and tools found in the field, allowing archaeologists to say where the obsidian of any particular artifact came from. The vast majority

FIGURE 55. Obsidian Butte, south end of Salton Sea.

of obsidian found throughout Southern California north of Imperial County comes from the Coso source in the Owens Valley.

If you drive up the 395, after passing Ridgecrest, you enter the southern end of the Owens Valley. You pass Little Lake, the road cuts through the western foot of a red cinder cone, and you're surrounded on all sides by lava fields. When you get to Coso Junction, there's a rest stop, and a little to the east there is the Coso Junction gas station and convenience store. If you point your vehicle east and look to the skyline in the distance, you'll see a massive gray-white mountainous dome of obsidian, the Sugarloaf obsidian source.

It's on naval weapons center land, and you can't drive in, but you can see it from the Coso Junction store parking lot, and it's mind-boggling to contemplate that millions of people over thousands of years traded for that obsidian because of its high quality and accessibility. Imagine the trade networks from present-day Inyo County crisscrossing all Southern California.

I had the privilege of working at the foot of Sugarloaf once, in the 1980s. David Whitley had a CRM job there and invited me to join him; he had a key to the gate. As we got closer and closer to the foot of the mountain, the obsidian flakes grew thicker and thicker underfoot. Each one of them was handmade, the result of a skilled procurement specialist, assaying and testing obsidian cobbles to determine the quality and nature of the glass, shaping the chunks down to carrying size that would fit in his rabbit-skin pouch or deer-skin backpack.

By the time we got to the base of the mountain, the flakes were crunching underfoot, as if God had dumped bags of black glass potato chips all over the ground—the flakes were that thick. Whitley had the idea to try to count them. We laid out a grid, I forget how many meters square, and counted the flakes inside the grid, then he multiplied that by the number of square grids it would take to cover the entire talus slope surrounding the butte and came up with a number in the billions! Each one made by hand by a live person, extracting resources for trade or personal use.

But for Imperial County, eastern San Diego County, and northern Baja California, Obsidian Butte by the Salton Sea was closer and more convenient, and archaeologists who have examined artifacts from sites in that region have found almost exclusive use of obsidian from this smaller yet nearer source.

The interesting thing about the Obsidian Butte source, and why I mention it in this chapter about Lake Cahuilla, is that when Lake Cahuilla was full,

Obsidian Butte was underwater. The people had to trade from other sources, including sources in today's Baja California, and of course the aforementioned Coso source. The preponderance of Obsidian Butte artifacts and flakes throughout Imperial County sites suggests that the lake was dry a lot more than it was full. When the lake was dry, the obsidian was available and obtainable, and its widespread appearance across the landscape probably means the lake was dry most of the time.

Another interesting cultural and historical fact about Obsidian Butte is its recent appearance on the landscape. Apparently, the liquid black glass flowed out of the ground and solidified around 2,500 years ago, which means that any artifact made from Obsidian Butte obsidian is automatically younger than 2,500 years. One more dating tool in the archaeologist's tool kit.

Tufa. Tufa, as mentioned, is the calcium carbonate deposit that accumulated around the edges of the ancient lake and coated the rocks that were in the wave zone for extended periods near the shore. Some locals call it coral because it has that rough and holey texture. I'm assuming that's how Coral Mountain in La Quinta got its name. A unique feature about tufa is that it lends itself perfectly to engraving or carving, easily chippable, and thus there are places around the old lakeshore with remarkable historic records: Cahuilla petroglyphs alongside 19th- and early 20th-century non-Native names and dates, ranchers that left their marks on the rocks.

For example, at one place near the western end of Avenue 64, there are tufa-coated rocks that bear ancient petroglyphs, including two stick figures, one with upraised arms and one with a crown with emanating rays. Alongside the Indian carvings we find "C. E. S. 1883," a very early date for non-Native settlement on the Coachella Valley, but why am I assuming it's non-Native? It could have been someone from nearby Toro village that had been to school and learned the Western alphabet and calendar. Tufa makes a good, permanent blackboard.

Once, on a job near Coral Mountain, a local old-timer told me he had seen initials carved in the "coral" where the Lewis and Clark expedition had come through. So much for local knowledge: only 1,000 miles off!

Pumice. Many of us are familiar with pumice stones used to scrub hard-water rings in toilet bowls, a very light rock formed when volcanic magma ejects under specific temperatures and pressures to create bubbles trapped within cooling rock. It has a foamy, sponge-like appearance, so light that it floats on water; in fact, Indians collected it where it washed up, wind driven,

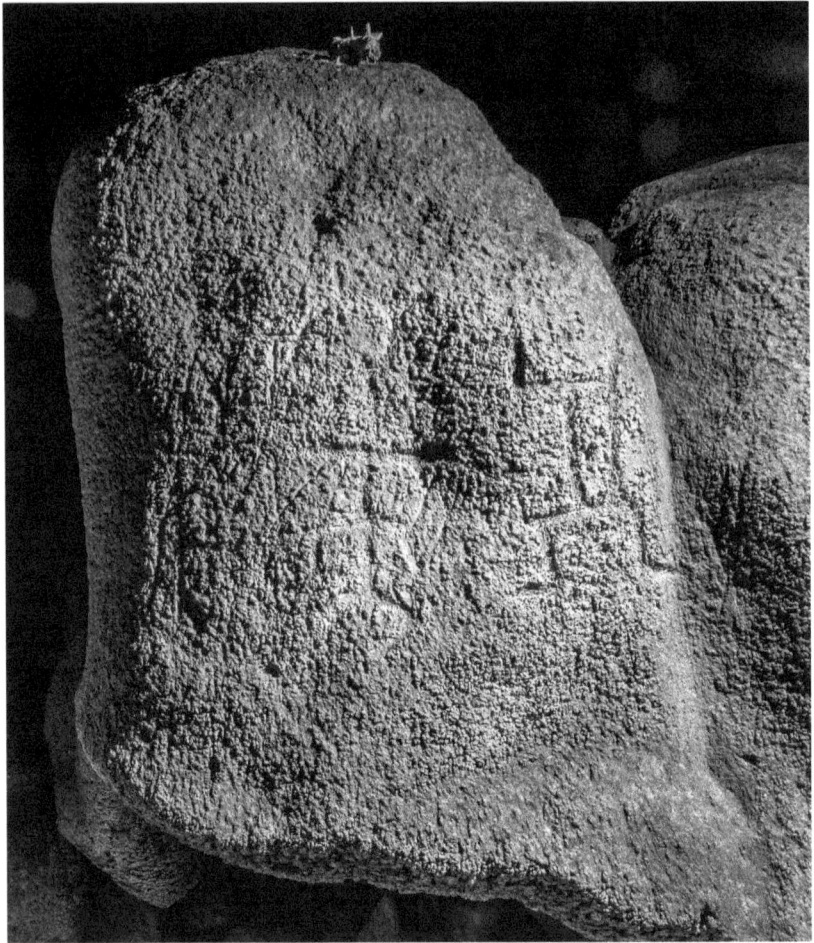

FIGURE 56. Cahuilla petroglyphs carved into tufa on boulders once underwater during a high lake stand, near west end of Avenue 64.

on the shores of ancient Lake Cahuilla, having come from the volcanic fields to the south around Obsidian Butte where the pumice deposits occur. Pieces of pumice are found throughout the archaeological deposits of the Coachella Valley, once used as hand-held abraders, scrapers, and arrow shaft straighteners. Artifacts made of pumice are yet one more aspect of Lake Cahuilla adaptation, one more piece in the cultural mosaic known as Indian lifeways in the Coachella Valley that is slowly being cobbled together through ethnography, ethnohistory, and archaeology.

FIGURE 57. Historical initials and date, "C. E. S. 1883," near west end of Avenue 64.

Anodonta. One last point worth mentioning regarding Lake Cahuilla: When scrapers and dozers cut swaths though the old lake beds during grading operations for future housing tracts, golf courses, and businesses, they expose layers of geological history as well as human history. One detail I found fascinating was that old lake surfaces are sometimes marked by siphon holes of freshwater mussels, anodonta.

You've maybe walked on an ocean beach when a wave washes up around your feet and then recedes and you see air bubbles coming up to the surface through holes in the saturated sand. Those bubbles mark the locations of marine clams. We found similar evidence of this phenomenon, only from freshwater mussels, on more than one job in La Quinta, in the vertical sidewalls of bulldozer cuts.

If you clean the surface carefully with your trowel, and you know what to look for, you might find remnant anodonta siphon holes, all in rows marking different elevations of shoreline surfaces. I credit the late Claude Warren, that doyen of desert archaeology, for pointing these out to me. Not necessarily archaeological—in other words not necessarily associated with artifacts or other evidence of human activity—but still, history.

Figure 63. Vertical exposure below Site CA-RIV-1974. Top of exposure is 35 ft above sea level. Note meter stick standing at base of cut. View is to the west, toward La Quinta and the Santa Rosa Mountains.

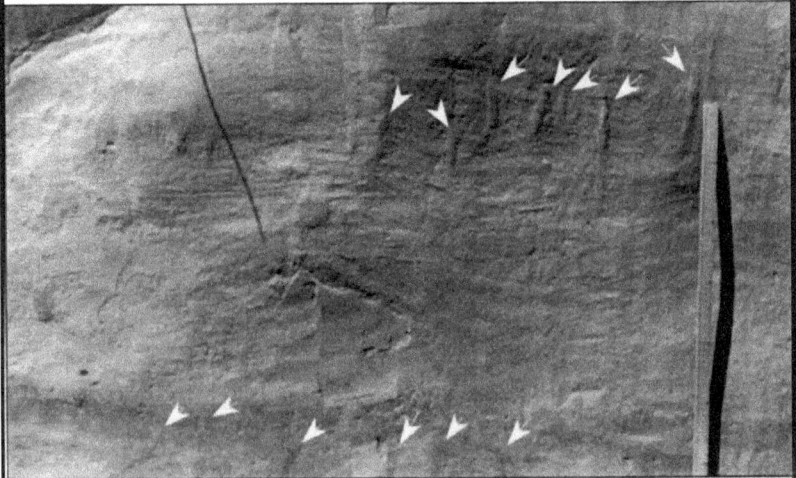

Figure 64. Fresh water mussel siphons leave traces in old lake shore sands. Arrows mark tops of siphon holes and old beach surfaces at 32 ft and 30 ft above sea level.

FIGURE 58. A page from one of our old reports (2001) showing freshwater mussel siphon holes that mark ancient lake levels.

SUMMARY

Why This Book?

T HANK YOU, DEAR reader, for coming with me on this personal retrospective. We working archaeologists tend to forget why we have these laws and requirements that protect archaeological sites and that require developers to jump through hoops to get clearance on their properties. Often we get lumped in with environmentalists because we are one more box to check when a developer applies for permits. After all, it is CEQA, the California Environmental Quality Act, that requires cultural resource studies on land development.

But the reason cultural resources, or archaeology, got included in these protections in the first place is because of their inherent value. No one has to convince anyone how important an Indian cemetery or an old adobe stagecoach stop is. When it came time for the legislature to pass CEQA, cultural resources were a no-brainer. No developer (well, almost no developer) wants to destroy something of true historic value. Virtually all the public, no matter what political leanings—left, right, center—speaks respectfully about history and even more respectfully about ancient history.

As working archaeologists, we need to remember this. The reason we get these jobs, these contract jobs, is because the laws were put in place for the public good. The problem is the yawning gap between what we do and the public's knowledge of what we do. CRM archaeologists go from job to job finishing reports that get filed away in city or county planning departments and at the statewide information centers.

At the annual Society for California Archaeology meetings and their off-shoots, the annual data sharing meetings, valuable information is shared and formal papers are presented. Academic articles get published in journals, but they rarely make it out to the general public; only the occasional newspaper article seems to cover them.

My own 13 years of CRM in the Coachella Valley (1990–2003) is a prime example. We accomplished so much, but so little is known about it: we pushed back the earliest dates of Indigenous occupation more than 2,000 years; recorded new arrowhead types and shell bead types; found clusters of fish traps previously unknown; explored, recorded, and tested a previously unknown field of bighorn sheep cairns; exposed burned remains of a Native house with a man and woman and their sacred possessions; saved dozens if not hundreds of cremations and returned them to their rightful owners, the nearby tribes; documented lakeshore villages and fish-roasting pits that confirmed a 17th-century high stand of ancient Lake Cahuilla; and so on.

Thus, this book.

Now, more than 20 years after leaving Riverside County, I find myself reflecting on those years and yearning to tell people what we did. The system *can* work: land developers, government planners, Native Americans, and archaeologists can all cooperate for the common good, but if the archaeological findings don't make it to the public consciousness, we've only done half our job. We collect and preserve the data, but who are the synthesizers that will translate technical reports into meaningful essays and articles for public consumption? In that vein, I present to you this book.

www.ingramcontent.com/pod-product-compliance
Lightning Source LLC
Chambersburg PA
CBHW040906210326
41597CB00029B/4993